MUTATION AS CELLULAR PROCESS

MUTATION AS CELLULAR PROCESS

A Ciba Foundation Symposium

Edited by

G. E. W. WOLSTENHOLME

and

MAEVE O'CONNOR

J. & A. CHURCHILL LTD.

104 GLOUCESTER PLACE, LONDON

1969

First published 1969

With 48 illustrations

Standard Book Number 7000 1430 6

Printed in Great Britain

Contents

Membership

Symposium on Mutation as Cellular Process held 11th–13th February, 1969

R. F. Kimball (Chairman)	Biology Division, Oak Ridge National Laboratory, Oak Ridge, Tennessee
D. Apirion	Department of Microbiology, Washington University School of Medicine, St. Louis, Missouri
Charlotte Auerbach	Institute of Animal Genetics, Edinburgh
H. Böhme	Institut für Kulturpflanzenforschung, Deutsche Akademie der Wissenschaften zu Berlin, Gatersleben
B. A. Bridges	MRC Radiobiological Research Unit, Harwell, Didcot, Berkshire
P. Brookes	Chester Beatty Research Institute, Pollards Wood Research Station, Chalfont St. Giles, Buckinghamshire
D. M. Brown	Department of Organic Chemistry, University Chemical Laboratory, Cambridge
H. Chantrenne	Laboratoire de Chimie Biologique, Faculté des Sciences, Université Libre de Bruxelles
C. H. Clarke	School of Biological Sciences, University of East Anglia, Norwich
G. W. P. Dawson	Department of Genetics, Trinity College, University of Dublin
R. Devoret	Centre des Faibles Radioactivités du CNRS, Gif-sur-Yvette
H. J. Evans	Department of Genetics, Marischal College, University of Aberdeen
L. Grossman	Graduate Department of Biochemistry, Brandeis University, Waltham, Massachusetts
R. Hütter	Mikrobiologisches Institut, Eidg. Technische Hochschule, Zürich
R. W. Kaplan	Institut für Mikrobiologie, Johann Wolfgang Goethe-Universität, Frankfurt am Main
B. J. Kilbey	Institute of Animal Genetics, Edinburgh
H. E. Kubitschek	Division of Biological and Medical Research, Argonne National Laboratory, Argonne, Illinois
A. Loveless	Chester Beatty Research Institute, London

O. Maaløe	University Institute of Microbiology, Copenhagen
G. E. Magni	Istituto di Genetica, Università di Milano
M. R. Pollock	Department of Molecular Biology, University of Edinburgh
A. Rörsch	Medical Biological Laboratory, The National Research Council, TNO, Rijswijk
W. L. Russell	Biology Division, Oak Ridge National Laboratory, Oak Ridge, Tennessee
F. H. Sobels	Department of Radiation Genetics, State University of Leiden
A. Wacker	Institut für Therapeutische Biochemie, Universität Frankfurt am Main
*Evelyn M. Witkin	Department of Medicine, Downstate Medical Center, State University of New York, Brooklyn, New York

* Contributed paper *in absentia*

The Ciba Foundation

The Ciba Foundation was opened in 1949 to promote international cooperation in medical and chemical research. It owes its existence to the generosity of CIBA Ltd, Basle, who, recognizing the obstacles to scientific communication created by war, man's natural secretiveness, disciplinary divisions, academic prejudices, distance, and differences of language, decided to set up a philanthropic institution whose aim would be to overcome such barriers. London was chosen as its site for reasons dictated by the special advantages of English charitable trust law (ensuring the independence of its actions), as well as those of language and geography.

The Foundation's house at 41 Portland Place, London, has become well known to workers in many fields of science. Every year the Foundation organizes six to ten three-day symposia and three to four shorter study groups, all of which are published in book form. Many other scientific meetings are held, organized either by the Foundation or by other groups in need of a meeting place. Accommodation is also provided for scientists visiting London, whether or not they are attending a meeting in the house.

The Foundation's many activities are controlled by a small group of distinguished trustees. Within the general framework of biological science, interpreted in its broadest sense, these activities are well summed up by the motto of the Ciba Foundation: *Consocient Gentes*—let the peoples come together.

Preface

This meeting was held as a result of Professor Charlotte Auerbach's suggestion, first made to Mr. A. V. S. de Reuck in 1966, that molecular biologists, geneticists and others would benefit from an exchange of information about the secondary processes involved in mutation. Whether a mutagen acts on a given gene and whether the product of this reaction develops into a mutation clone or organism depends largely on cellular metabolism. Yet workers on mutation were only beginning to realize the importance of cellular events for the mutation process, and people studying cellular processes did not always appreciate the importance of their findings for mutation research. The symposium recorded here therefore brought representatives of the relevant different disciplines together to attempt to analyse the problem of mutation and reduce it to simpler levels.

Professor Auerbach gave much invaluable advice throughout the organization of the meeting and her help at every point was greatly appreciated. Dr. Kimball, who acted as Chairman, carried out this task with apparent ease, tact and excellent time-keeping, even though he had to cope with several last-minute changes in the programme. These changes arose because the blizzard which hit the East Coast of the United States in February 1969 prevented three members of the symposium from reaching London at the time expected. Two, Dr. L. Grossman and Dr. W. L. Russell, arrived after journeys of epic proportions, but after three frustrating days Dr. Evelyn Witkin finally had to give up her strenuous attempts to get to the meeting.

The editors wish also to thank Dr. A. Loveless for his generous help during the early stages of publication.

CHAIRMAN'S OPENING REMARKS

R. F. KIMBALL

As you all know, this symposium was initiated by Professor Charlotte Auerbach. Its subject reflects her view that mutation is not just an isolated event, a quantum event, or a simple chemical reaction, but a process in which cellular functions are intimately involved. This view I share, and therefore it is a pleasure and an honour for me to be asked to chair this symposium.

This is not to say that quantum events and simple chemical reactions are not the starting points of processes leading to mutation. Most of us, however, work with cellular systems and here we cannot forget that the mutagenic treatments we give initiate a process which only after an appreciable time results in our detection of mutations. Cellular metabolism and cellular events are intimately involved in this process.

The two points of view, mutation as a simple molecular event and as a more complex cellular process, have been in existence for a long time and attention has shifted back and forth between them. Concentration on the initial atomic and molecular event leads to a certain simplicity that is especially attractive to those who come to the problem from physics and chemistry. Concentration on the cellular process aspect is more congenial to those of us who come to the problem from biology and are more immediately aware of the complexities of biological systems. Here we must be very careful, however, that this awareness of complexity is not converted to a worship of it and a consequent resistance to the attempt to analyse the problem and reduce it to simpler levels. It is just this analysis and reduction that should be one of the main purposes of this symposium.

Let me start with a very crude overall view of mutation as a cellular process. We can recognize at least the following main steps: (1) initiation, (2) fixation, (3) detection. With many mutagens, initiation necessitates molecular alteration of the chromosome. In general, however, this alteration—an alkylated base for example—will not itself be self-replicating but must be converted to some self-replicating change, such as a base-pair transition. There are many possible variations of these first two steps. In some variations, such as a spontaneous mistake by the replicating system, the two steps might be reduced to one, but with most induced mutations initiation and fixation are certainly separate. We have learned much in the last few years about the molecular aspects of those processes that can eliminate initial damage before it can be fixed, but we still understand all too little about the molecular aspects of fixation itself. Much evidence and some theory suggest that fixation normally occurs within the same cell cycle in which the initial event occurred, often associated with DNA replication, but apparent replicating instabilities suggest that sometimes fixation may be delayed for several or many cell generations.

By detection, I mean all those events, processes, and procedures that are involved in detecting a mutation once it has been fixed. This requires among other things that the mutant gene be expressed, i.e. that it make its gene product. In modern terminology transcription and translation must occur. Even if this happens, however, we must be able to recognize the effects of the gene product by some *test*, which may depend for its success on cellular metabolism. Among other things, a successful test may require that the mutant gene be *segregated* free of non-mutant ones. Techniques for doing this with germ cells have been part of the standard methods of genetics for many years. With single cells, one has to be aware that the chromosome consists of more than one conserved strand and that mutations may be fixed in only one strand. Segregation will be further complicated if fixation can be delayed for several cell generations.

Because mutation is a cellular process, modifications of mutation yield are not necessarily the result of direct action on the mutation process *per se*, but may be the result of actions upon the cell that affect mutation yield quite indirectly. As a result, the interpretation of many experiments is made difficult or ambiguous. For example, a modifying treatment given before fixation or even before the initiation might modify the post-fixation process of expression as a consequence of an alteration in cellular metabolism or in the cell cycle. The time of treatment is not by itself a safe guide to the portion of the mutation process that is affected.

Despite the difficulties caused by such complexities, much progress has been made as better and better methods for analysing the molecular events in initiation and fixation have been developed and a greater understanding has been reached of the processes involved in detection. I hope that this symposium will help to accelerate this progress by bringing together people with divergent interests who can look at the mutation process from different points of view.

INFLUENCE OF THE HOST ON THE INDUCTION AND EXPRESSION OF MUTATIONS IN PHAGE KAPPA

R. W. Kaplan, R. von Lohr and M. Brendel

Institut für Mikrobiologie der Universität Frankfurt am Main

The process of induced mutation consists of several steps or phases, each of which could be influenced by cell components in the vicinity of the mutating gene. One way to study such influences is to use phage mutations which allow that the phage can be treated, e.g. irradiated, separately from the host cell. This has been done for ultraviolet-induced and to a smaller extent with X-ray-induced mutations in phage κ.

This temperate phage forms turbid plaques on a (red) lawn of *Serratia marcescens*, strain HY. Several types of plaque mutations appear after irradiation of the free phage (Kaplan, Beckmann and Rüger, 1963; Beckmann and Kaplan, 1965). U.v. induces mainly clear-plaque (*c*) types to the extent of a few per cent among survivors. The dose/mutation curve indicates a two-hit process (Ellmauer and Kaplan, 1959). A large fraction of these mutations is photoreversible (Winkler, 1965*c*). X-rays induce *c*-mutations and also the rare type *b* by a one-hit process, and type *e* by about three hits (Rüger and Kaplan, 1966). Only *c*-mutations will be considered here because they are the easiest to distinguish from the turbid wild-type plaques; they map in four regions of the κ-genome. U.v.-induced *c*-types are mainly mutated in region III which is responsible for the phage repressor (Steiger, 1966, 1968). The following problems have been studied: (1) The influence of irradiation of the host protoplasm on mutation induction in the phage genome by u.v. and X-rays, in particular the "indirect action" of the radiation; (2) The influence of the growth phase of the host; (3) The dependence of the expression of the mutations as pure or mosaic clones on host cell reactivation (HCR).

INFLUENCE OF IRRADIATION OF THE HOST CELL

Winkler (1963) found that u.v. irradiation of the host cells (HY) alone does not cause mutations in unirradiated phage κ. Therefore, mutagenic products of u.v. in the host protoplasm stable enough to survive until infection cannot be responsible for the mutations observed in the phage.

In order to see whether u.v. produces short-lived mutagens, e.g. free radicals, within the cell which could induce mutations in the injected phage-DNA, we irradiated the phage following infection of the host cell. Since it was necessary to inhibit intracellular phage development until the end of irradiation, cells in stationary phase were used. They were starved in saline for 60 minutes and then allowed to adsorb unirradiated phage for 15 minutes; after washing they were irradiated with different doses of u.v. (mainly 254 nm, 20 erg mm^{-2} sec^{-1}). The results show (Fig. 1) that the yield of *c*-mutations as well as the inactivation was much higher with intracellular than with extracellular irradiation. The dose reduction was of the order of 2, perhaps for mutation a little higher. This increase could be due to the hypothetical short-lived u.v.-mutagens. However Winkler (1963) had already observed that u.v. irradiation of the host cells before infection increased the frequency of *c*-mutations in u.v.-irradiated phage by a factor up to 2 as compared to infection of unirradiated host cells. Thus, the increase observed with intracellular irradiation could also be due to this effect since the host cell had simultaneously received u.v.

To decide between these alternatives the effect of intracellular irradiation was compared with the results of an experiment in which both the free phage and the host received the same u.v. doses before infection. This was done with host cells starved as before. They were irradiated quickly immediately after addition of phage to the cells and before adsorption had become significant. This technique ensured that the intensity of u.v. irradiation of the phage (which was lower due to shielding by the cells) was nearly the same as with intracellular irradiation. The results show that the yield of mutations as well as the inactivation is about the same as with intracellular irradiation (Fig. 1, points X). From this it can be concluded that the effect of intracellular irradiation is not more than the sum of the u.v. effect on the free phage plus that due to irradiation of the uninfected cells. Thus, neither short nor long-lived mutagenic radiation products from the host protoplasm are markedly involved in u.v. mutation of this phage.

The same was found to be true for X-rays. The hypothesis that X-rays induce mutations indirectly, via radicals etc., has long been discussed. We (Rüger and Kaplan, 1966) had already shown that an "indirect" effect of X-ray products of water did not significantly induce mutations in extracellular phage κ as the mutation frequency was the same in the presence and absence of protecting substances such as broth. We therefore X-irradiated intracellular phage and found the mutation frequency to be no

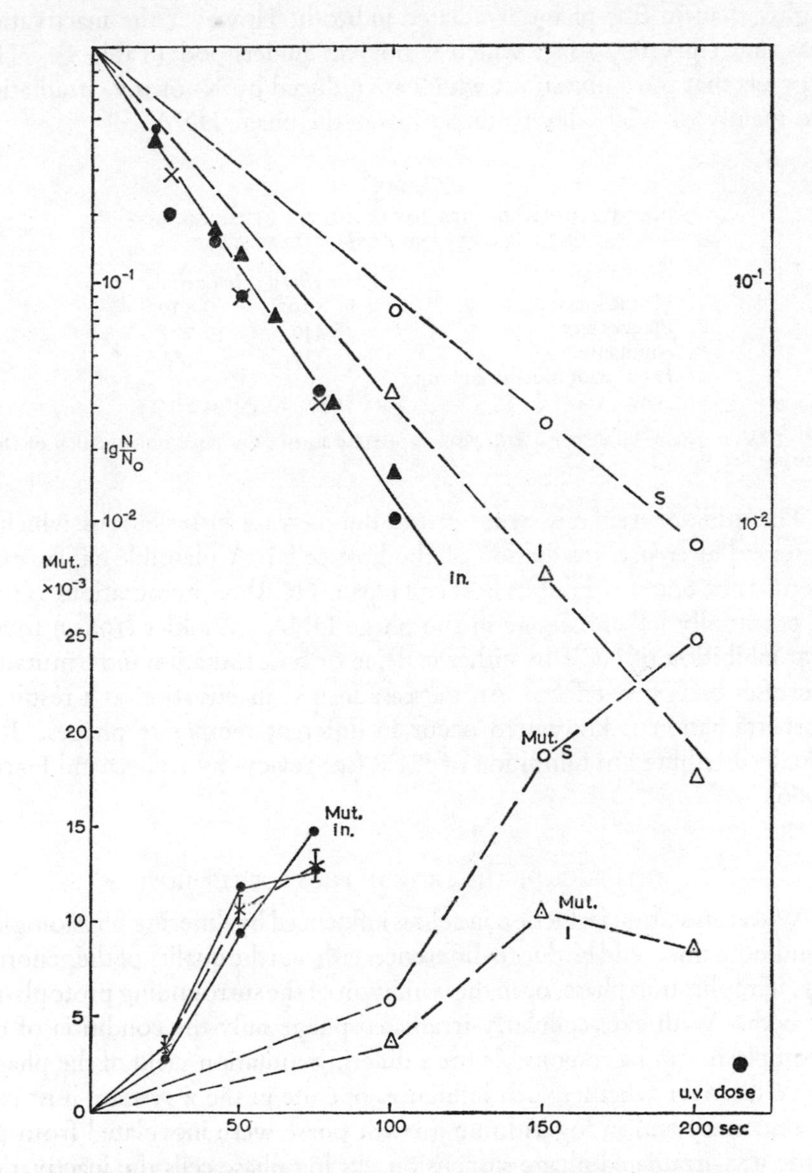

FIG. 1. Survival (lg N/N₀) and frequency of *c*-mutations in phage ϰ as a function of the u.v. dose. Irradiation of extracellular (○, △) or intracellular (in. ●, ▲) phage using stationary (s; ○, ●) or logarithmic growing (l; △, ▲) cells of strain HY of *Serratia marcescens* as host for preadsorption. X = Irradiation of free ϰ and stationary host cells with the same doses before infection. All platings with HY. Stationary cells were grown overnight in Difco Nutrient Broth (NB) and resuspended in buffered saline. Log cells were grown overnight in NB, then diluted (10^{-2}) into fresh NB, grown for 2½ hours, resuspended in buffered saline and used immediately. All incubations at 30°C.

higher than in free phage irradiated in broth. However the inactivation was much greater, a fact which is not yet understood (Table I). This suggests that the *c*-mutations which are induced by X- or u.v. irradiation are mainly or solely due to direct hits in the phage DNA.

TABLE I

INACTIVATION AND MUTATION OF κ-PHAGE AFTER X-IRRADIATION* IN EXTRA- AND INTRACELLULAR STATES

	Extracellular	*Intracellular*
Plaque survival	$2 \cdot 8 \times 10^{-1}$	3×10^{-4}
Plaques scored	18 440	57 769
c-mutants	16	54
Fraction of mutants among survivors ($\times 10^{-3}$)	$0 \cdot 87 \pm 0 \cdot 22$	$0 \cdot 94 \pm 0 \cdot 13$

* 55 kV; 0·3 mm Al; dose 234 krd; phage suspended in 10 times usual concentration of Difco Nutrient Broth.

The question remains, what causes the increase in mutations which is observed after u.v. irradiation of the host cells? A plausible explanation seems to be that u.v. inhibits host cell repair (HCR) of premutations as well as potentially lethal damage in the phage DNA. Winkler (1965*a*) found that inhibition of HCR by either caffeine or host mutation (*hcr*⁻ mutants) increases both u.v. effects. An increase in u.v. inactivation as a result of host-irradiation is known to occur in different temperate phages. It is usually attributed to inhibition of HCR (see review by Rupert and Harm, 1966).

INFLUENCE OF THE GROWTH PHASE OF THE HOST

Where mutation induction in cells is influenced by differing physiological conditions this could be due to differences either in the quality of the genome, e.g. its replication phase, or in the condition of the surrounding protoplasm, or both. With extracellularly-irradiated phage only the condition of the protoplasm can be responsible for a differing mutation yield of the phage.

To discover whether such influences operate in the κ system, host cells in stationary and in logarithmic growth phase were inoculated from the same u.v.-irradiated phage suspension. In log phase cells the inactivation was greater than in stationary cells but the mutation yield was significantly lower (Fig. 1). For the same extent of inactivation, 1·4 times the dose is required when the host cells are in stationary phase as when they are in log phase, but for mutation the factor is only about 0·6 or 0·8.

It is interesting that this influence of growth phase on inactivation disappeared with intracellular irradiation (Fig. 1). This may be a hint that

HCR is involved. It seems reasonable to assume that HCR can act on the phage DNA for a longer time in resting than in log phase cells since the latent period of phage production was found to be 80 minutes in stationary cells, but only 65 minutes in log cells. Also u.v. is much less effective in preventing colony formation when irradiation is given to stationary phase cells, and the survival curve has a much broader shoulder.

To test the influence of HCR stationary and log phase cultures of the mutant *hcr*$^-$42 (Winkler, 1965*a*) were used as hosts. The irradiated phage was preadsorbed, the remaining free phage removed by washing, and the complexes were plated on *hcr*$^+$ strain HY. No significant difference in phage survival was observed. Indeed, in some of the experiments a very small reverse difference was found (Fig. 2). On the other hand, the mutation induction was again much lower in log than in stationary host cells. When *hcr*$^-$42 and HY, both in stationary phase, were compared as hosts, the extent of inactivation and the yield of mutants was much smaller for HY. This agrees with the earlier findings of Winkler (1965*a*). The dose reduction factors are 12 to 15 for inactivation and about 3 for mutation.

What could be the reason for the much lower mutation induction in log as compared to stationary host cells? Since HCR is excluded two possibilities may be considered:

(1) To produce a mutation from a premutation in the DNA, e.g. from a pyrimidine dimer, an error in replication has to occur. The conditions in log cells may provide less chance for such errors to occur than those in stationary cells, e.g. they may influence the exactness of action of the replicase.

(2) As in other cases of differing sensitivity to radiation, a repair mechanism could be involved. This repair must of course be different from HCR, it must act preferentially on premutations as compared with lethal lesions, and it must be more active in log than in stationary cells.

The second hypothesis of a "log phase mutation repair" is perhaps more plausible than the first because it could explain the weak reactivation of plaque formation observed in some of the experiments with *hcr*$^-$42 when there is also a weak repair of otherwise lethal damage.

A type of repair acting mainly on premutations in phage κ was discovered by Winkler (1965*b*) in his two u.v.-sensitive mutants, *hcr* 91 and 614, and by Steiger and Kaplan (1964) in the *Serratia* strain CN. None of these hosts gives any mutations with u.v.-irradiated κ-phage, although they plate "old" *c*-mutants as clear plaques very well. They also do not allow X-ray-induced mutations of κ (Winkler, 1965*b*; Rüger and Kaplan, 1966). This mechanism for "extinguishing" all u.v. and X-ray-induced premutations in κ phage resembles the EXR mechanism of *E. coli*. As Witkin (1967) found, the *exr*⁻ strains of this bacterium do not yield u.v. mutations in the bacterial genome. She proposed that post-replication gaps in the DNA strand opposite to the premutated strand are closed later by repair replication. This produces errors, i.e. mutations, in *exr*⁺ strains but no errors, i.e. wild-type, in *exr*⁻. Perhaps the non-mutating strains contain an apparatus reversing different types of premutationally changed DNA bases (e.g. pyrimidine dimers as well as certain X-ray products) to the wild-type state.

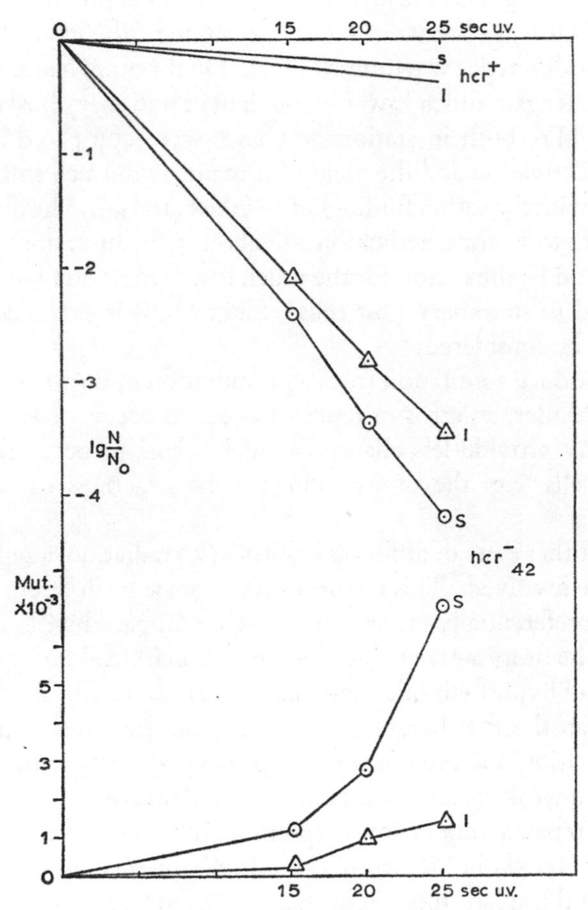

FIG. 2. Survival (lg N/N_0) and frequency of *c*-mutations in phage κ as a function of u.v. dose. Irradiation of extracellular κ, preadsorption for 15 min on host strain *hcr*⁻42 in stationary (s) or logarithmic growth (l) phase, then washing to remove free phage and plating with strain HY. Above: survival in strain HY (*hcr*⁺) for comparison.

INFLUENCE OF HCR ON U.V.-INDUCED MOSAICS AND PURE MUTANT CLONES

It was reported earlier (Kaplan, 1966) that most u.v.-induced *c*-plaques of κ contain only mutant phages. Several hypotheses exist (listed e.g. by Nasim and Auerbach, 1967) to explain the finding of pure mutant clones from two-stranded DNA in different organisms: (1) the premutation is located in both strands; (2) a lethal hit in the non-premutated strand inhibits the replication of this strand alone; (3) a repair replication in the strand opposite to the premutation transfers the mutation to this strand too ("mutating repair"); (4) one strand only, the master strand, gives the information to both daughter DNA molecules ("simplex" hypothesis of Barricelli, 1965).

The lethal-hit hypothesis predicts an increase in the fraction of pure clones among all clones containing mutants with increase in the u.v. dose, since relatively more strands with such hits would occur at higher doses. A weak increase was indeed observed earlier in u.v.-induced *c*-mutation of phage κ (Kaplan, 1966). Therefore, a "purification" of the mutant clones could be caused, at least partially, by such hits. But mutating repair could also be involved since the host strain HY used at that time has a repair potential (HCR). A case where mutating repair seems to play a role was demonstrated in hydroxylamine-induced mutation in phage T4 by Bautz Freese and Freese (1966).

To test the influence of HCR for the κ system the clone composition of u.v.-induced *c*-mutations was studied using the hosts *hcr*⁻42 and HY (*hcr*⁺). Phage κ was u.v. irradiated and preadsorbed onto each strain. Both types of complexes were then plated on HY and many of the *c*-plaques thus obtained were picked, and the virus was diluted and replated. The results (Table II) show that the defect of HCR in *hcr*⁻42 compared with

TABLE II

PURE AND MOSAIC *C*-MUTANT PLAQUES INDUCED BY U.V.

No. of expts.	U.v. -dose (erg mm^{-2})	Survival	c-plaques picked	Mosaics* (with wild type among c-plaques)
(A) Host strain HY (*hcr*⁺), plated on HY				
2	4800	10^{-4}	331	6 (1·8%)†
2	3600	10^{-3}	236	8 (3·4%)
(1966)	5400	10^{-4}	94	1 (1·6%)
(1966)	3600	10^{-3}	95	9 (9·5%)
(B) Host strain *hcr*⁻42, plated on HY				
4	600	10^{-5}	95	13 (13·7%)
3	350	10^{-3}	110	7 (6·4%)

* Wild types and mutants of all mosaics listed were stable in later post-cultures.

† Besides the six mosaics with wild type, four plaques contained two stable mutant types: three with about 0·1% *c*-type besides *l*-type and one with 82% *c*- and 18% *l*-type.

hcr$^+$ has increased the proportion of mosaics among all *c*-plaques from 1·8 per cent to 13·7 per cent at the high doses, and from 3·4 to 6·4 per cent at the low doses. The latter difference is not significant but the first is ($P = 1 \times 10^{-5}$).

The influence of the dose in *hcr*$^+$ was smaller in this new series than in the earlier one, and alone was not significant, but the pooled results of both series show statistical significance ($P = 0{\cdot}008$). In *hcr*$^-$42 the dose effect even seems to be reversed but this is not significant ($P = 0{\cdot}11$). When the distribution of the clone types (percentage of mutants in a clone) was considered the sample seemed to have a deficit of clones with few (< 50 per cent) mutants. Special experiments with mixed infections (by 50 per cent wild types plus 50 per cent of 30 u.v.-induced *c*-types) showed that clones with few mutants often produce plaques looking like wild type. Therefore in the new u.v. series many plaques appearing to be wild type were picked and replated. Several clones were found which contained few mutants but they were only a minority of those appearing as *c*-plaques. The clone distribution obtained in these experiments was added to those from *c*-plaques in order to obtain a corrected distribution (Fig. 3).

A comparison of the distributions obtained with the two hosts shows that active HCR is correlated with a relative increase of completely pure clones (with 100 per cent mutants) from 45 per cent in *hcr*$^-$42 to 81 per cent in *hcr*$^+$ ($P = 10^{-7}$ for the difference). This increase is paralleled by a decrease in clones with few wild phages (99 to 50 per cent mutants): 33 per cent in *hcr*$^-$42 and 4 per cent in *hcr*$^+$ ($P = 3 \times 10^{-3}$). One can assume that both these clone types descended from premutated phage DNA of which the first daughter DNA was mutated, either by mutating repair or by the master strand mechanism. The wild-type genomes in the nearly pure (99 to 50 per cent mutant) clones would be produced mostly in later generations, e.g. by replication of the still premutated parental DNA without a copy error or by such DNA where the premutation had later been removed by repair. Since HCR converts nearly pure to completely pure clones it seems that it is acting as a mutating repair mainly in late generations of the vegetative phage. This interpretation is supported by the result (Fig. 3) that more clones with only few mutants (< 6 per cent) were found in *hcr*$^+$ (12 per cent) than in *hcr*$^-$ (0 per cent), though the difference is not significant ($P = 0{\cdot}06$). Such clones could arise when mutating repair again yields mutant DNA *late* in the growth of the vegetative phage clone which consists in this case mainly of wild type.

The conclusions are that HCR is not responsible for the predominance of pure or nearly pure mutant clones induced by u.v. in phage κ. It seems

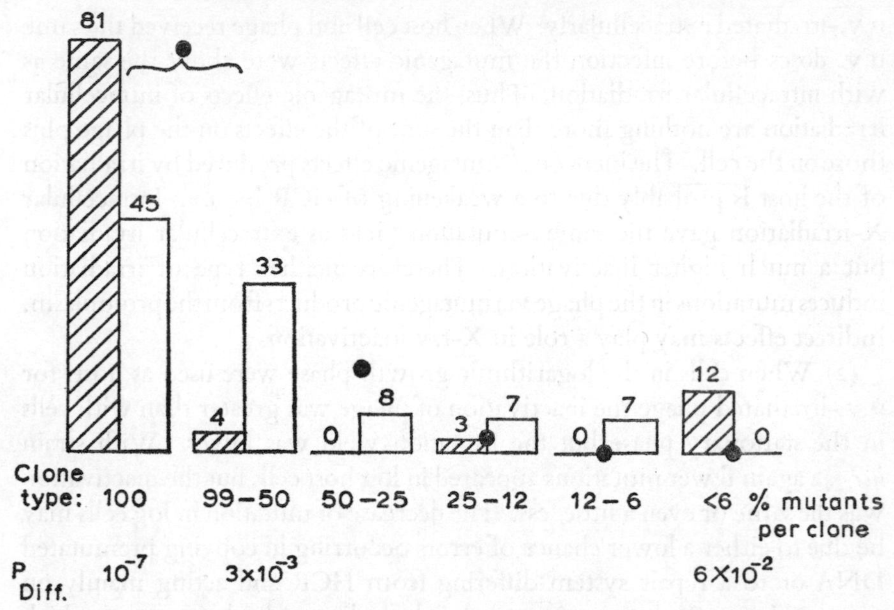

FIG. 3. Distribution (%) of frequencies of clones with various percentages of mutants (clone types) after u.v.-irradiation of phage ϰ, using strains HY (*hcr*+) or *hcr*−42 as preadsorption hosts and plating on HY.

U.v. doses: 4800 erg mm^{-2} (survival 3×10^{-4}) for HY; 600 erg mm^{-2} (survival 3×10^{-5}) for *hcr*−42.

▨ in *hcr*+

□ in *hcr*−42

●: Frequencies expected according to the simplex model (Barricelli, 1965) with M=0·8.

P_{Diff}: Probability of obtaining the difference observed between *hcr*+ and *hcr*− by random variation only.

to be only an additional "purifying" factor, acting mainly late in clone growth as a mutating repair mechanism. Its main effect is to "extinguish" part of the u.v. premutations. The main cause of the clone purity could be a different type of mutating repair, e.g. the one indicated by the non-mutating strains or by the effect of log growth phase. But the master strand ("simplex") mechanism could also be responsible. If the formula given for this hypothesis by Barricelli (1965) is used for calculating the expected clone-type distribution, the data obtained with *hcr*−42 fit the hypothesis, assuming a copying error chance of M=0·8 (Fig. 3). For *hcr*+ an additional effect of HCR, e.g. the one proposed, has to be assumed to fit the results better.

SUMMARY

(1) When phage ϰ was u.v.-irradiated inside the host cell both inactivation and the yield of *c*-mutations were higher than with phage

u.v.-irradiated extracellularly. When host cell and phage received the same u.v. doses before infection the mutagenic effects were about the same as with intracellular irradiation. Thus, the mutagenic effects of intracellular irradiation are nothing more than the sum of the effects on the phage plus those on the cell. The increase in mutagenic effects produced by irradiation of the host is probably due to a weakening of HCR by u.v. Intracellular X-irradiation gave the same *c*-mutation yield as extracellular irradiation but a much higher inactivation. Therefore neither type of irradiation induces mutations in the phage via mutagenic products from the protoplasm. Indirect effects may play a role in X-ray inactivation.

(2) When cells in the logarithmic growth phase were used as hosts for u.v.-irradiated phage the inactivation of phage was greater than with cells in the stationary phase but the mutation yield was lower. With strain *hcr*⁻42 again fewer mutations appeared in log host cells but the inactivation was the same or even a little less. The decrease of mutation in log cells may be due to either a lower chance of errors occurring in copying premutated DNA or to a repair system differing from HCR and acting mainly on premutations. Such a repair was already indicated by host strains which "extinguish" all u.v.- as well as X-ray-induced premutations.

(3) Most u.v.-induced *c*-mutant plaques contained only *c*-phages. The small proportion which had some wild type as well as mutant type was a little higher at a lower u.v. dose. Thus, the "purification" of mutant clones may only partly be due to recessive lethal hits. The proportion of mixed clones was also slightly higher with the host $hcr^{-}42$ than with hcr^{+}. The distribution of clone types (assessed by picking *c*- as well as wild-type plaques) fits the "simplex" model of Barricelli in $hcr^{-}42$. Thus, HCR is not the main cause of clone purity. In hcr^{+} an additional mechanism producing mutant genomes instead of wild-type seems to be present, which may be due to (mutating) HCR acting late in intracellular clone multiplication.

Acknowledgements

This work was supported by the Deutsche Forschungsgemeinschaft by a grant to the first author. We thank Mrs. H. Stoye for carefully performing many of the experiments.

REFERENCES

BARRICELLI, N. A. (1965). *Virology*, **27**, 630–633.
BAUTZ FREESE, E., and FREESE, E. (1966). *Genetics, Princeton*, **54**, 1055–67.
BECKMANN, H., and KAPLAN, R. W. (1965). *Z. allg. Mikrobiol.*, **5**, 1–18.
ELLMAUER, H., and KAPLAN, R. W. (1959). *Naturwissenschaften*, **46**, 150.
KAPLAN, R. W. (1966). *Photochem. Photobiol.*, **5**, 261–264.

KAPLAN, R. W., BECKMANN, H., and RÜGER, W. (1963). *Nature, Lond.*, **199**, 932–33.
NASIM, A., and AUERBACH, C. (1967). *Mutation Res.*, **4**, 1–14.
RÜGER, W., and KAPLAN, R. W. (1966). *Z. allg. Mikrobiol.*, **6**, 253–269.
RUPERT, C. S., and HARM, W. (1966). *Adv. Radiat. Biol.*, **2**, 1–81.
STEIGER, H. (1966). *Z. VererbLehre*, **98**, 111–126.
STEIGER, H. (1968). *Molec. gen. Genet.*, **103**, 21–28.
STEIGER, H., and KAPLAN, R. W. (1964). *Z. allg. Mikrobiol.*, **4**, 367–389.
WINKLER, U. (1963). *Z. Naturf.*, **18b**, 118–123.
WINKLER, U. (1965*a*). *Z. VererbLehre*, **97**, 18–28.
WINKLER, U. (1965*b*). *Z. VererbLehre*, **97**, 29–39.
WINKLER, U. (1965*c*). *Z. VererbLehre*, **97**, 75–78.
WITKIN, E. M. (1967). *Brookhaven Symp. Biol.*, **20**, 17–55.

DISCUSSION

Clarke: In stationary phase and log phase hosts is the mutational yield after u.v. proportional to the number of rounds of mating of phage ϰ within these hosts? Is the burst size different for stationary and log phase cells?

Kaplan: Phage ϰ has one round of mating in the hcr^+ strain compared to five for T4 and 0·5 for λ, but the value is not known for hcr^-. I think strain *hcr* 42 is a true hcr^-: it behaves like such strains of *E. coli*.

Kimball: Have these strains been tested for their ability to excise thymine dimers?

Kaplan: No.

Bridges: You suggested that the reason why phage are more sensitive to both the lethal and mutagenic actions of u.v. when irradiated intracellularly is because host-cell repair is inhibited by u.v. Therefore if cells are being killed, for example, lesions are being put into the phage genome and concomitantly the repair capacity of the cells is being reduced. This should give you a continuous downward curvature for the survival curve, which in fact you don't see.

Kaplan: Not for the phage, but we see it for colony formation.

Bridges: But you should get it for the phage too. And with mutations you should get an upward curvature to the dose-response curve for intracellularly irradiated phage but not for phage irradiated extracellularly. Do you get any difference in the shape of the response curve?

Kaplan: Are you suggesting that when we give a high dose to the host we should get a single-hit curve for inactivation and a shoulder with the non-irradiated host?

Bridges: The other way round. If you haven't irradiated the host then you should get a single-hit survival curve whatever type of mutation curve you get, whether it is one-hit or two-hit. If you then irradiate the host, as you would do if you were irradiating the complex, you should get superimposed on these curves an increasing curvature downwards on the survival curve and upwards on the mutation curve because you are inhibiting repair as you increase the dose of u.v. So you should get different shapes, not just different slopes, and this didn't seem to be so in your data.

Kaplan: When the HCR system is completely inactivated, as happens at quite low doses, then we should not get this curvature. The doses given to the phage (to inactivate or mutate it) and equally to the host are very high for the host. They all saturate HCR.

Maaløe: To what extent could the loss of phage after u.v. irradiation be due to faulty injection? Your argument breaks down unless 100 per cent of the infecting phage particles inject this DNA.

Kaplan: We have a selective system in this phage and have found that cross-reactivation occurs. This means that phage DNA is injected even at rather high doses. The adsorption of irradiated phage to the host is normal.

Maaløe: Experiments by French and co-workers (1951) with ^{32}P-labelled phage indicated that in the act of injection degradation of the DNA can occur, with the label spilling out into the medium. If the killing that you observe is to some degree caused by this kind of misadventure it becomes very difficult to superimpose effects on the extracellular and on the already injected material.

Kaplan: All I want to say is that with intracellular irradiation there is no effect additional to the effect on the irradiated host alone.

Evans: At a similar mutation rate level of around 1×10^{-3} with both X-rays and u.v. light, your data reveal a 1000-fold difference in survival levels. Is this large order of difference between the killing potential of these two radiations also evident at other mutation rate levels?

Kaplan: I don't know; higher doses are difficult to investigate because of the strong killing of complexes.

Evans: Could the difference in lethality that you observe be due to a saturation of the HCR system at the high dose levels used?

Kaplan: We have not yet X-irradiated the host cell alone, so we cannot see whether X-rays inhibit HCR in our system. Our main interest was in the mutation induction.

Loveless: I would have thought that, except for 100 per cent mutant bursts, an analysis of plaque content was an extremely unreliable indication of the population of the first burst on a plate. Have you any evidence on this matter, or have you any direct evidence on the population of your first bursts as analysed by single-burst experiments?

Kaplan: We have not yet done single-burst experiments. In mixed infection by wild and *c*-type there was no indication that selection influenced the growth within the plaque.

Loveless: If you give mixed infections of mutant and wild type do they emerge in the first burst in the same proportion as you have put them in? If they don't, then your mixed infection experiments have no relevance.

Kaplan: The proportion of wild to *c*-type in the mixed infection was 50:50. There was no marked difference in the plaques.

Auerbach: Mosaics may be of two entirely different kinds. First, there may be a mutation in only one strand of DNA; this will usually lead to a large admixture of wild-type phage. Second, there are replicating instabilities; these may give rise to clones with very few mutants. The issue becomes confused quantitatively if one pools the two types.

Kaplan: This is an important point. There are two kinds of mixtures. In one an unstable allele may be produced which would mutate spontaneously during the growth of the clone. In the other, stable types are produced which give a kind of segregation in the clone.

Auerbach: In the organisms with which I am familiar the distinction is fairly easy. With phage one would have to see whether one and the same wild-type phage particle can give rise to *several* mutated ones, in which case it is an instability that can replicate as such.

Kaplan: We made further post-cultures of the first progeny plates and noted whether plaques of wild type and of *c*-type were stable or not. We found only one case of instability in further post-cultures. All the cases shown in Table II (p. 9) are stable. We tested all types of clones. In the first series we only picked the *c*-plaques. But there was the mentioned deficit of clones with much wild type which was attributed to the fact that clones with an excess of wild type often look like wild-type plaques. Therefore a great number of wild-type plaques appearing after u.v. irradiation were picked. They contained in fact a few clones with a minority of mutants. We added these values to the distribution of the first series of pickings of *c*-plaques and got the distribution I showed in Fig. 3 (p. 11). Plaques with 40–50 per cent *c*-type look clearer than wild-type plaques.

Evans: In Table II, Professor Kaplan, you showed that the percentage of mosaics decreased with increasing mutation frequency or increasing dose. This in fact agrees with one of the four hypotheses put forward to account for the origin of mosaics, as listed by Nasim and Auerbach (1967). Is the reverse of this result seen in *Drosophila* sperm in situations where there is no in-built replicating instability, i.e. is there any evidence for an increasing percentage of mosaics with increasing dose?

Sobels: In *Drosophila* the ratio of "completes" to mosaics increases with increase of X-ray dose. It is also higher with other dose-modifying factors, such as X-irradiation in oxygen versus that in nitrogen, or irradiation of spermatids compared to that of sperm (Inagaki and Nakao, 1966). The *Drosophila* data available would therefore tend to support the lethal-hit hypothesis.

Maaløe: Professor Kaplan, your observation that exponentially growing cells gave quite different results to resting cells when infected with u.v.-irradiated phage κ might be easier to understand if we knew more about the experimental procedure. Did you score by plating infected cells or was lysis allowed to occur? Did you take stationary phase cells and dilute them into the medium in which they are infected? In the latter case the cells would be pulling out of the stationary phase during infection. In a rapidly growing cell, a characteristic sequence of synthesis leads to lysis at an early time with release of many complete phage particles. In the resting phase cell several copies of the phage genome may have to wait around for an abnormally long time before enough capsid material and enough lysozyme are produced. Such a differential effect seems likely because a resting phase cell has very few ribosomes. The net result might be a difference in the average number of rounds of mating in the pool of DNA, which could account for the high level of mutants; recombination is assumed to indicate "fixation" of the mutant genotype.

Kaplan: The irradiated phage was used to infect cells of an overnight culture in broth (called stationary culture) after the cells had been washed. The same phage was used to infect cells that had grown for a few hours in broth (called log cells). The multiplicity was about 0·1. Then these cells infected in different growth states were plated, before lysis, with the same indicator culture. Probably the intracellular development of the phage under the two cultural conditions is different, as already indicated by the different periods of latency. This difference in the response of the growing phage population according to the conditions within the host, for instance to repair enzymes, may have produced the difference found in the mutation rate and inactivation.

REFERENCES

French, R. C., Lesley, S. M., Graham, A. F., and van Rooyen, C. E. (1951). *Can. J. med. Sci.*, **29**, 144–148.

Inagaki, E., and Nakao, Y. (1966). *Mutation Res.*, **3**, 268–272.

Nasim, A., and Auerbach, C. (1967). *Mutation Res.*, **4**, 1–14.

INFLUENCE OF CELLULAR PHYSIOLOGY ON THE REALIZATION OF MUTATIONS—RESULTS AND PROSPECTS

C. H. Clarke

School of Biological Sciences, University of East Anglia, Norwich

Two main approaches have been used to analyse the steps in the mutational pathway (Clarke, 1967*b*). These may be called the physiological and the genetic approaches. In the first, one studies the effects of such influences as growth conditions, growth factors, analogues, repair inhibitors, temperature changes, antibiotics or photoreactivating light upon the production of a stable base-pair change in the DNA and its subsequent final phenotypic expression. The second, genetic, approach involves the use of mutants lacking one or more repair functions or having other types of altered phenotype.

From the range of potential mutational systems, both forward and reverse, available in even one organism, and the range of potential mutagens it should be obvious that there is no one unique mutational pathway. We are concerned to know, therefore, whether the results found for one particular mutational system, e.g. $str^s \rightarrow str^r$, with one particular strain of bacterium, e.g. *Escherichia coli* B/r, grown under one particular set of conditions, e.g. glucose-starved stationary phase cells grown in minimal medium, and treated with one particular mutagen, e.g. 253·7 nm ultraviolet light, apply only to these conditions or are of more general interest and validity (Shankel, 1962). It is one of the tragedies of mutational research up to the present that so little careful comparative work has been done. The numerous non-agreements in the literature probably therefore merely reflect the fact that different workers have used different systems and methods to study mutagenesis. Some of these differences have already been pointed out (Clarke, 1967*a*). In view of this it would be well to be extremely cautious of generalizations made from any one set of results. Much of the analysis of the mutational pathway in bacteria has been carried out with *E. coli* strain B/r, using ultraviolet light or ionizing radiations. Much less work has been done with other mutagens, other strains of *E. coli*, or other enteric bacteria, let alone other Gram-negative or Gram-positive bacteria or eukaryotic organisms. There have been numerous, perhaps too numerous, purely physiological studies on mutagenesis. With the introduction of genetic approaches, however, the purely physiological results can, in some cases, now be interpreted with some real degree of understanding (Witkin, 1966*a*, *b*, 1967).

As examples of physiological influences on mutagenesis for which we still have no reasonable biochemical explanations I might mention the findings of Haas and Doudney (1957). They found that pre-incubation of *E. coli* B in the presence of *p*-aminobenzoic acid or riboflavin, before exposure to u.v., led to an increased yield of eosin-methylene blue (EMB) colour variants. A similar effect was produced by pre-feeding with RNA ribosides.

An example of a physiological effect on mutagenesis for which a satisfactory biochemical explanation is probably near is the inhibitory action of L-methionine on the phenotypic expression of *adn*$^+$ reversions in the fission yeast, *Schizosaccharomyces pombe* (Clarke, 1962, 1963). L-Methionine was found to exert a whole variety of effects on *adn*$^-$ cells, all probably attributable to *S*-adenosylmethionine accumulation (Clarke, 1965).

In the genetic sphere one may mention the long-known influences of the genetic background on mutability. These include mutator genes, enhancers and diminishers of specific mutations or mutagens, and—at present receiving great attention—repair gene mutants. In only some of these cases do we yet know the underlying modes of action of these genic effects on mutagenesis. For example, the disappearance of u.v. revertability of the ochre *try*$^-$ mutant, WP-2, in *E. coli* B/r on introduction of a *str*-d marker into the strain (Witkin and Theil, 1960) is probably explained on the basis of lowered efficiency of an ochre suppressor in a cell with altered ribosomes (Gartner and Orias, 1966).

The greatest progress in analysing the steps in the mutational pathway has come recently from the use of repairless mutants (Witkin, 1966*a*, *b*, 1967; Hill, 1965; Munson and Bridges, 1966; Bridges, Dennis and Munson, 1967; Bridges and Munson, 1968). This in turn has led to an understanding of the action of some physiological agents such as white light, temperature changes, caffeine and acriflavine in terms of repair inhibition. We thus have evidence that the enzymic photoreactivation system is specific for pyrimidine dimers, and that such photoreactivable dimers most certainly give rise to some, but not all, mutations in *E. coli* B/r (Bridges, Dennis and Munson, 1967; Witkin, Sicurella and Bennett, 1963; Witkin, 1964). There is evidence that excision repair can deal with damage caused by mitomycin

C, nitrous acid and mustards, but not with damage caused by nitrosoguanidine (Witkin, 1967). Repair of X-ray and methyl methanesulphonate (MMS) damage may well share some steps in common with the excision repair of ultraviolet damage (Reiter *et al.*, 1967; Zimmermann, 1968). It seems likely that caffeine, directly or indirectly, inhibits the activity of the nickase, the first enzyme in excision repair (Doneson and Shankel, 1964; Shankel and Kleinberg, 1967). There is no marked caffeine effect upon X-ray, diepoxybutane or nitrosomethylurethane mutagenesis in *E. coli* B/r (Clarke, 1969). However, repair systems may exist for some types of chemical damage which have no action on u.v. damage. Strauss and Robbins (1968) have already described an enzyme specific for methylated (MMS-treated) but not ultraviolet-irradiated DNA. Very speculatively I would suggest that manganous chloride might inhibit the dark-repair of some types of chemical damage, without acting on other repair systems (Böhme, 1962; Arditti and Sermonti, 1962).

MUTATION FREQUENCY DECLINE

I should like to deal in some detail with the so-called mutation frequency decline (MFD) dark-repair system (Doudney and Haas, 1958; Witkin, 1966*a*, *b*). This exists in *E. coli* B/r and acts on u.v.-induced potentially mutagenic lesions, its action being inhibited by an amino acid pool. The MFD system is of particular interest because of the specificities and peculiarities which have been attributed to it. Firstly, MFD is held to be a dark-repair system acting, in broth-grown cells, best in late lag phase and only weakly in exponential or stationary phase cells (Witkin, 1966*b*; Clarke, 1967*b*). Secondly, MFD is said to be without effect upon potentially lethal lesions, or premutations at streptomycin resistance loci, and to be specific for premutational lesions at supersuppressor loci. Thirdly, MFD is said to be due to excision repair acting preferentially on only some u.v.-induced lesions in the bacterial genome (Witkin, 1966*a*, *b*). Fourthly, MFD is without action upon γ-induced damage (Munson and Bridges, 1966) but there is some evidence that it may affect nitrous acid premutational lesions (Rudner, 1961).

I shall examine these assumptions critically, and describe experimental results which bear on the interpretation of MFD. MFD was invoked to explain the differences in *try*$^+$ revertant yields which were obtained, in u.v.-irradiated *Salmonella typhimurium* and *E. coli* B/r tryptophan auxotrophs, between platings on broth- (or amino-acid-pool-) enriched as against minimal plus low level tryptophan medium. It is as well to point out immediately that the *S. typhimurium* and *E. coli* B/r systems do not show identical characteristics. Broth-grown stationary, and probably other, phase cells of *S. typhimurium try* C-3 when u.v.-irradiated give high frequencies of *try*$^+$ reversions as long as platings are made on minimal medium plus broth, or minimal plus a low level of tryptophan plus an amino acid pool. Very low frequencies of *try*$^+$ reversions are obtained when platings are made either on unsupplemented minimal or minimal plus a low level of tryptophan (Witkin, 1956; P. H. Williams, 1969, personal communication). *S. typhimurium try* C-3 cells grown in tryptophan-supplemented minimal medium before u.v. irradiation gave high frequencies of revertants on minimal plus low level tryptophan even without an additional amino acid pool, i.e. no broth effect (Witkin, 1956).

In *E. coli* B/r WP-2 *try*$^-$ grown in nutrient broth the enhancement of revertant frequencies by an amino acid pool, over and above any effect due to tryptophan alone, applied to late lag phase cells and not nearly as strongly to stationary phase cells (Clarke, 1967*a*, *b*). Furthermore WP-2 cells grown in tryptophan-supplemented minimal medium show an enhancement of *try*$^+$ frequencies by broth over and above any effect due to tryptophan (Munson and Bridges, 1966). The differences between the results with *S. typhimurium* and *E. coli* B/r might lie in different enzymes being affected by mutation in the two strains. Thus only some of the enzymes concerned in trytophan synthesis contain tryptophan residues in their amino acid sequences (Yanofsky and Ito, 1966). Alternatively the *S. typhimurium*—*E. coli* differences may reflect differing mechanisms for control of the transcription of transfer RNA loci.

MFD was invoked to explain the enhancing effect of a non-specific amino acid pool on revertant yield. Thus onc criterion of MFD is a large excess of revertants appearing on plates of minimal medium plus a low, priming, level of the specific requirement, plus an amino acid pool, as compared with the same medium lacking only the unspecific amino acid pool. It is important to note that a comparison between totally unsupplemented minimal medium and minimal plus a low level of the specific requirement plus an amino acid pool is invalid and misleading (Kada, Brun and Marcovich, 1960; Strauss and Okubo, 1960).

MFD is also indicated by the occurrence, on incubation of u.v.-irradiated auxotrophs in unsupplemented liquid minimal medium, of an irreversible decrease in the number of revertants subsequently scorable on broth- (or amino-acid-pool-) supplemented plates. MFD has been followed in liquid complete, i.e. amino-acid-rich, or on solid complete media containing chloramphenicol. Under such conditions revertant numbers decrease in spite of the presence of the amino acid pool. Chloramphenicol thus abolishes the inhibition of MFD caused by the amino acids.

One would like to know whether the three criteria for MFD in fact all measure the same process. Could one, for example, obtain a large broth effect on plates of supplemented minimal medium while yet finding no rapid decrease in mutant numbers on incubation in liquid minimal medium? Can one *validly* compare MFD results obtained by different techniques? There are some results which suggest that one may not be able to do so. Firstly, by the chloramphenicol-nutrient agar plate technique Witkin and Theil (1960) found that u.v.-induced *str*^{r} and *str*^{ind} mutations are not susceptible to MFD. However, using different test conditions, Shankel (1962) and Shankel and Coupe (1962) found that *str*^{r} mutations do show MFD and are inhibited by chloramphenicol treatment. Secondly, Doudney and Haas (1959) showed that for the *E. coli* B/r *try*$^{-}$ auxotroph WP-2 the addition of L-tryptophan, the specific requirement, alone markedly inhibited the rate of MFD in liquid minimal medium.

If MFD were in fact excision repair (Witkin, 1966*a*, *b*) one might expect MFD to act not absolutely exclusively upon mutational, but also, to some extent at least, upon lethal lesions. *Hcr*$^{-}$ strains, unable to carry out excision repair, have been shown by Witkin (1966*a*, *b*) and Munson and Bridges (1966) to exhibit greatly reduced rates of MFD in liquid minimal medium. However, as pointed out by Witkin (1966*a*), this does not prove that MFD is excision repair. In the *hcr*$^{-}$ strains the premutational lesions are mostly genuine enzymically photoreactivable dimers, whereas in the *hcr*$^{+}$ strains this is not the case.

Mfd^{-} mutants have been isolated by Witkin and these show an absence of MFD in liquid minimal medium. I have compared one of these mutants, 36–10–45 (Witkin, 1966*a*), with the parent strain 36–10 which is *mfd*$^{+}$. Strain 36–10 is a B/r diauxotroph requiring tyrosine and leucine. The tyrosine marker, WU-36, is an ochre mutation and the leucine marker, 10, is an amber (Osborn and Person, 1967; Clarke, 1967*b*).

Broth-grown late lag phase cells of 36–10 and 36–10–45 have been irradiated with ultraviolet light, then plated for both survival and reversion estimates immediately after irradiation and also after known periods of incubation in liquid minimal medium. Survival and *tyr*$^{+}$ reversions have been scored on minimal *plus* a high level (20 μg/ml) of L-leucine *plus* a low level (1 μg/ml) of L-tyrosine and also on minimal *plus* high level leucine *plus* 2·5 per cent, by volume, of Oxoid nutrient broth. Likewise both survival and *leu*$^{+}$ reversions have been scored on minimal *plus* high level tyrosine *plus* low level leucine and also minimal *plus* high level tyrosine *plus* broth (Clarke, 1967*b*). Survival has thus been scored on four media, two with and two without broth (=amino acid pool), and *leu*+ and *tyr*+ reversions on two media each. Three separate experiments were performed with each strain.

The results obtained are shown in Figs. 1, 2 and 3. Fig. 1 represents a plot of revertant *numbers* against time of post-irradiation incubation in minimal medium. There is a classical MFD effect, with the *leu*+ type, scored on broth-enriched medium, decreasing before the *tyr*+ type of revertant

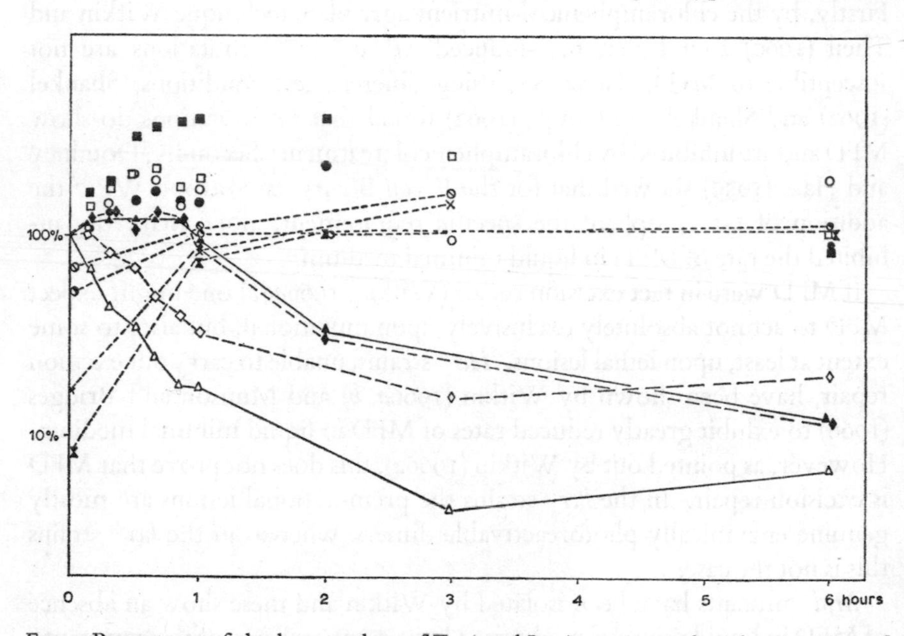

FIG. 1. Percentages of absolute numbers of Tyr+ and Leu+ revertants plotted against time of post-irradiation incubation in liquid minimal medium. Late lag phase broth-grown cells of *E. coli* B/r *tyr*−*leu*− diauxotrophs 36–10 (*mfd*+) and 36–10–45 (*mfd*−). Experiments carried out using the methods of Witkin (1966*a*) and Hill (1968).

Open symbols for strain 36–10, filled symbols for strain 36–10–45:

●, ○ = minimal + high level tyrosine + low level leucine } Leu+ reversions
——▲, ——△ = minimal + high level tyrosine + broth } Leu+ reversions
■, □ = minimal + high level leucine + low level tyrosine } Tyr+ reversions
----◆, ----◇ = minimal + high level leucine + broth } Tyr+ reversions

100 % = No. of revertants appearing on plates of any one medium on plating immediately after irradiation:

○ 100 % = 34	● 100 % = 399
△ 100 % = 851	▲ 100 % = 2931
□ 100 % = 69	■ 100 % = 244
◇ 100 % = 495	◆ 100 % = 2831

----× = survival on broth-free media, strain 36–10
----⧗ = survival on broth-free media, strain 36–10–45
----⊘ = survival on broth-enriched media, strain 36–10
----◑ = survival on broth-enriched media, strain 36–10–45

in strain 36–10. This result confirms the findings of Witkin (1966*a*) with the same strain, and includes the *leu*+ result also. With strain 36–10 the numbers of revertants of both the *tyr*+ and *leu*+ types appearing on broth-enriched plates fall rapidly to a low plateau level. On broth-free plates the numbers remain low throughout. In strain 36–10–45 similar plots of absolute numbers of revertants against time show delayed, but *not* totally absent MFD. It is important to emphasize that these post-irradiation incubations are carried out in unsupplemented liquid minimal medium without either tyrosine or leucine. It is clear from the results shown in

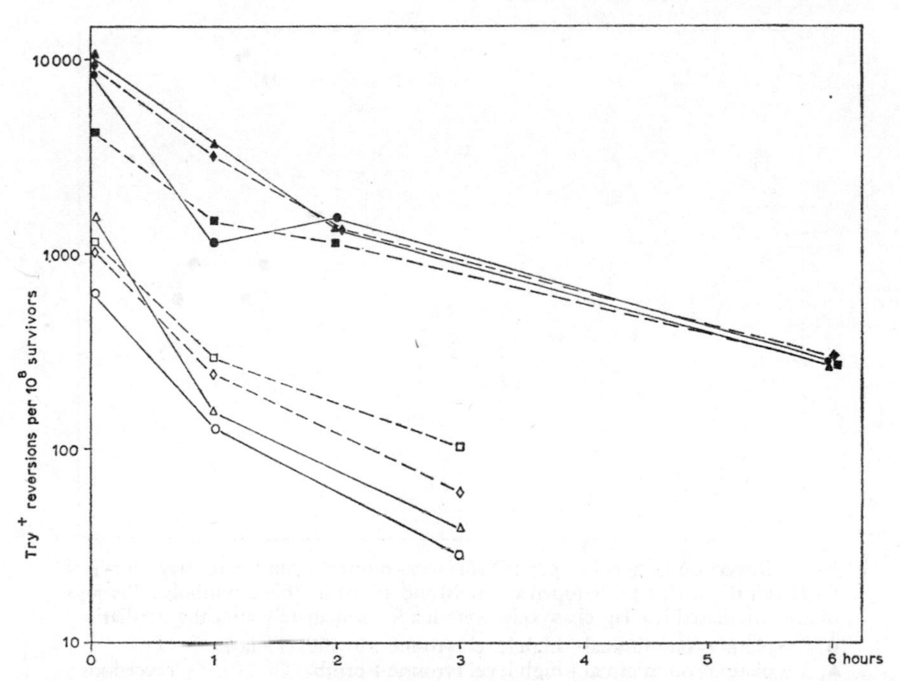

FIG. 2. Reversion frequencies per 10^8 survivors plotted against time of post-irradiation incubation in liquid minimal medium. Late lag phase cells of *E. coli* B/r strains 36–10 (open symbols) and 36–10–45 (filled symbols).

●, ○ = platings on minimal + high level tyrosine + low level leucine } Leu+ reversions
▲, △ = platings on minimal + high level tyrosine + broth

■, □ = platings on minimal + high level leucine + low level tyrosine } Tyr+ reversions
◆, ◇ = platings on minimal + high level leucine + broth

Fig. 1 that for survival estimates made immediately after u.v. irradiation, survival is much higher on broth-enriched than on broth-free plates. This applies to both strain 36–10 and strain 36–10–45. Furthermore during the course of incubation in liquid minimal medium (MFD conditions) survival increases. This result strongly supports the contention of Hill (1968) and

Clarke (1967*b*) that survival recovery, in at least a fraction of the u.v.-irradiated population, is a major factor in MFD. These ideas receive strong confirmatory support from the earlier results of Alper and Gillies (1960), Gillies and Brown (1967), Tabaczyński (1962), Okagaki (1960), and Kos, Drakulić and Brdar (1965) showing that factors which tend to promote the recovery of wild-type *E. coli* B/r, and similar strains, are identical to those factors promoting MFD in B/r auxotrophs. Thus some, though probably not all, of the enhancement of revertant frequencies by the amino acid

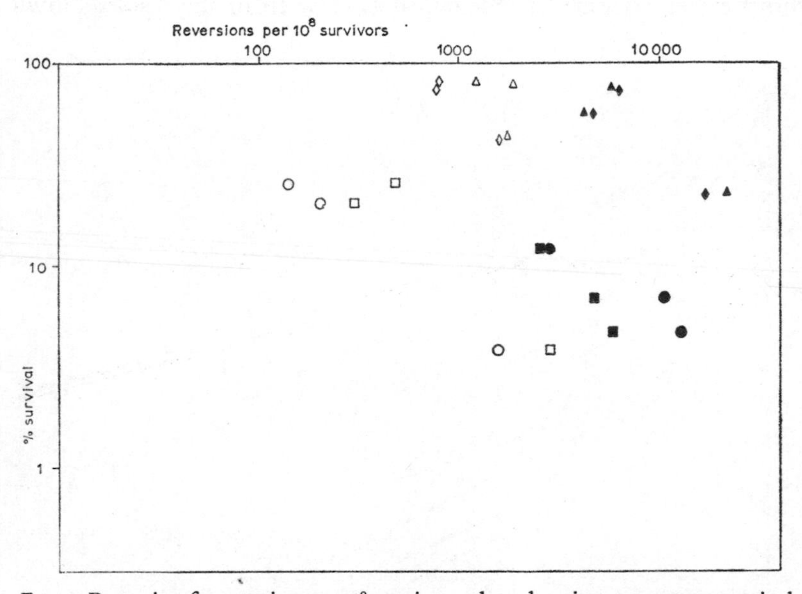

FIG. 3. Reversion frequencies per 10^8 survivors plotted against percentage survival for *E. coli* B/r strains 36–10 (open symbols) and 36–10–45 (filled symbols). Platings of u.v.-irradiated late lag phase cells were made immediately after the irradiation.

●, ○ = platings on minimal + high level tyrosine + low level leucine } Leu⁺ reversions
▲, △ = platings on minimal + high level tyrosine + broth
■, □ = platings on minimal + high level leucine + low level tyrosine } Tyr⁺ reversions
◆, ◇ = platings on minimal + high level leucine + broth

pool is due to differences in survival levels on the different plating media.

If, taking account of the different survivals on the different media, one plots not absolute numbers of revertants but rather frequencies of revertants per 10^8 survivors against incubation time in minimal medium, one obtains the results shown in Fig. 2. The curves in Fig. 2 are indicative of MFD for both broth-rich and broth-free platings. It is important to note that to a large extent the falls in revertant frequencies can be attributed merely to increased survival, i.e. effective dose-reduction. Failure to realize this in the past has, most probably, been due to the use of high initial survival levels, so that any subsequent increase would be small, and to the failure to use broth-enriched as well as broth-free plating media for survival estimates.

It is important to note that both 36–10 *and* 36–10–45 show the recovery of survival and decreased revertant frequencies on post-irradiation incubation in liquid minimal medium. In strain 36–10–45 the recovery process is merely delayed. Doudney and Haas (1958) showed that 5-hydroxyuridine causes a superficially similar delayed type of MFD, and as mentioned earlier the required amino acid was also shown to cause a retardation of MFD. Strain 36–10–45 may differ from 36–10 not in its excision repair ability but perhaps in the size of its amino acid pool, the extent of its lag phase, and the control of transcription at tRNA loci. Plots of revertant frequencies against percentage survival for the late lag phase cells of both 36–10 and 36–10–45 are informative (Fig. 3). Fig. 3 suggests that strain 36–10–45 is more mutable than 36–10, in the sense that at any given percentage survival level, on any one medium, reversions are more frequent in 36–10–45 than in 36–10.

Strains 36–10 and 36–10–45 do not differ in ultraviolet sensitivity (Witkin, 1966*a*, *b*; Clarke, unpublished results), so 36–10–45 is probably not deficient in excision repair. Furthermore the curves obtained by plotting revertant frequencies against survival levels (Fig. 3) are unlike those found for pairs of *hcr*$^+$ and *hcr*$^-$ strains (Witkin, 1966*b*). One further test for excision-repair differences between 36–10 and 36–10–45, which has not so far been applied, is to determine whether or not these two strains differ in the extent to which caffeine or acriflavine will enhance u.v.-induced killing and reversions.

It seems that the final nature of MFD and the amino-acid-pool effect is still unclear. Part of the phenomenon can certainly be explained on the basis of survival differences and changes, probably resulting from dark-repair activity. There remains the possibility that some degree of differential dark-repair at supersuppressor loci, perhaps associated with the transcription of tRNA loci, is involved in MFD. Further analysis of MFD should include experiments in strains, and under conditions, where survival differences and changes either do not occur or are in the opposite direction to those in *E. coli* B/r. *E. coli* strain B, and *exr*$^-$ and *rec*$^-$ mutants of B and B/r are good candidates in this respect. It should be obvious, but it has often been ignored, that incubation of auxotrophic cells in a minimal medium after, and even without, u.v. irradiation disturbs a great many biochemical processes (Neidhardt, 1966). Various agents, such as starvation for the required amino acid, 6-azauracil (Doudney and Haas, 1960), pyronin B

(Witkin, 1961) and chloramphenicol, have been found to have the same final effect of lowering revertant frequencies. One would like to know at a biochemical level the common factor(s) in their modes of action.

It seems important to establish clearly whether MFD applies to mutagens other than u.v., and whether it is absolutely specific for supersuppressor reversions. One would also like to know whether analogous systems exist in eukaryotic organisms. There is a need to examine, in some detail, *mfd*$^-$ mutants to know in what aspects of cellular physiology they differ from their *mfd*$^+$ parents, and to determine whether *mfd*$^-$ mutants map in known dark-repair genes.

Further analysis of the mutational pathway, other than solely at the level of repair systems, can come, I believe, from detailed biochemical analyses of genetic background effects. Perhaps, in practice, one ought to pay attention not only to mutation potentiators, such as excision repair-inhibitors, but also to antimutagens. Caffeine, under some conditions (Clarke, 1967*a*, 1968; Grigg and Stuckey, 1966), and polyamines (Johnson and Bach, 1966), are known examples. There is a need too to carry out u.v. and γ mutagenesis experiments with synchronized cultures of bacteria to determine whether or not these agents, like nitrosoguanidine (Cerdá-Olmedo, Hanawalt and Guerola, 1968), act preferentially at, or perhaps before or after, the DNA replication fork. There are some reasons for guessing that X- or γ-rays might be mutagenic only when acting on already-replicated genes. More information is needed on the factors which govern genic accessibility to mutagenesis and repair. The agent chloro-IPC (isopropyl-*N*-(3-chlorophenyl) carbamate) may perhaps prove useful in some of these experiments.

SUMMARY

Recent analysis of mutational pathways has come mainly from *E. coli* B/r, with ultraviolet light and ionizing radiations. The main concern has been with the activity and inhibition of repair systems. It seems likely that this work will now be extended to other, chemical, mutagens and to other systems. Evidence is presented, supporting that of Hill (1968), that one supposedly highly specific repair system, mutation frequency decline, may be largely a reflection of changes in survival level. Further analysis of the mutational pathway could well necessitate studies of repression, transcription, replication and translation involvement. Suggestions are made for experiments involving synchronized cultures, antimutagens, and the biochemical analysis of genetic background and physiological effects.

Acknowledgements

The work described in this paper has been supported by funds from the Medical Research Council and the University of East Anglia. I wish to acknowledge my debt to all my former colleagues at the M.R.C. Mutagenesis Research Unit, Edinburgh, friends in North America, and Dr Donald MacPhee in Norwich, for interesting and enthusiastic discussions and suggestions. My especial thanks go to Drs Ruth F. Hill, Evelyn Witkin and Bryn Bridges for so kindly providing bacterial strains and details of their own work and results before publication.

REFERENCES

ALPER, T., and GILLIES, N. E. (1960). *J. gen. Microbiol.*, **22**, 113–128.
ARDITTI, R. R., and SERMONTI, G. (1962). *Genetics, Princeton*, **47**, 761–768.
BÖHME, H. (1962). *Biol. Zbl.*, **81**, 267–276.
BRIDGES, B. A., DENNIS, R. E., and MUNSON, R. J. (1967). *Genetics, Princeton*, **57**, 897–908.
BRIDGES, B. A., and MUNSON, R. J. (1968). *Proc. R. Soc. B*, **171**, 213–226.
CERDÁ-OLMEDO, E., HANAWALT, P. C., and GUEROLA, N. (1968). *J. molec. Biol.*, **33**, 705–719.
CLARKE, C. H. (1962). *Z. VererbLehre*, **93**, 435–440.
CLARKE, C. H. (1963). *J. gen. Microbiol.*, **31**, 353–363.
CLARKE, C. H. (1965). *J. gen. Microbiol.*, **39**, 21–31.
CLARKE, C. H. (1967*a*). *Molec. gen. Genet.*, **99**, 97–108.
CLARKE, C. H. (1967*b*). *Molec. gen. Genet.*, **100**, 225–241.
CLARKE, C. H. (1968). *Mutation Res.*, **5**, 33–40.
CLARKE, C. H. (1969). *Mutation Res.*, **8**, 35–41.
DONESON, I. N., and SHANKEL, D. M. (1964). *J. Bact.*, **87**, 61–67.
DOUDNEY, C. O., and HAAS, F. L. (1958). *Proc. natn. Acad. Sci. U.S.A.*, **44**, 390–401.
DOUDNEY, C. O., and HAAS, F. L. (1959). *Proc. natn. Acad. Sci. U.S.A.*, **45**, 709–722.
DOUDNEY, C. O., and HAAS, F. L. (1960). *Genetics, Princeton*, **45**, 1481–1502.
GARTNER, T. K., and ORIAS, E. (1966). *J. Bact.*, **91**, 1021–1028.
GILLIES, N. E., and BROWN, D. (1967). *Biochem. biophys. Res. Commun.*, **26**, 102–107.
GRIGG, G. W., and STUCKEY, J. (1966). *Genetics, Princeton*, **53**, 823–834.
HAAS, F. L., and DOUDNEY, C. O. (1957). *Proc. natn. Acad. Sci. U.S.A.*, **43**, 871–883.
HILL, R. F. (1965). *Photochem. Photobiol.*, **4**, 563–568.
HILL, R. F. (1968). *J. gen. Microbiol.*, **52**, 261–270.
JOHNSON, H. G., and BACH, M. K. (1966). *Proc. natn. Acad. Sci. U.S.A.*, **55**, 1453–1456.
KADA, T., BRUN, E., and MARCOVICH, H. (1960). *Annls Inst. Pasteur, Paris*, **99**, 547–566.
KOS, E., DRAKULIĆ, M., and BRDAR, B. (1965). *Nature, Lond.*, **205**, 1125–1126.
MUNSON, R. J., and BRIDGES, B. A. (1966). *Mutation Res.*, **3**, 461–469.
NEIDHARDT, F. C. (1966). *Bact. Rev.*, **30**, 701–719.
OKAGAKI, H. (1960). *J. Bact.*, **79**, 277–291.
OSBORN, M., and PERSON, S. (1967). *Mutation Res.*, **4**, 504–507.
REITER, H., STRAUSS, B., ROBBINS, M., and MARONE, R. (1967). *J. Bact.*, **93**, 1056–1062.
RUDNER, R. (1961). *Z. VererbLehre*, **92**, 336–360.
SHANKEL, D. M. (1962). *J. Bact.*, **84**, 410–415.
SHANKEL, D. M., and COUPE, B. (1962). *Bact. Proc.*, G85.
SHANKEL, D. M., and KLEINBERG, J. A. (1967). *Genetics, Princeton*, **56**, 589.
STRAUSS, B., and OKUBO, S. (1960). *J. Bact.*, **79**, 464–473.
STRAUSS, B. S., and ROBBINS, M. (1968). *Biochim. biophys. Acta*, **161**, 68–75.
TABACZYŃSKI, M. (1962). *Acta microbiol. pol.*, **11**, 301–312.
WITKIN, E. M. (1956). *Cold Spring Harb. Symp. quant. Biol.*, **21**, 123–140.
WITKIN, E. M. (1961). *J. cell. comp. Physiol.*, **58**, 135–144.

WITKIN, E. M. (1964). *Mutation Res.*, **1**, 22–36.
WITKIN, E. M. (1966*a*). *Science*, **152**, 1345–1353.
WITKIN, E. M. (1966*b*). *Radiat. Res.*, suppl. 6, 30–53.
WITKIN, E. M. (1967). *Brookhaven Symp. Biol.*, **20**, 17–55.
WITKIN, E. M., SICURELLA, N. A., and BENNETT, G. M. (1963). *Proc. natn. Acad. Sci. U.S.A.*, **50**, 1055–1059.
WITKIN, E. M., and THEIL, E. C. (1960). *Proc. natn. Acad. Sci. U.S.A.*, **46**, 226–231.
YANOFSKY, C., and ITO, J. (1966). *J. molec. Biol.*, **21**, 313–334.
ZIMMERMANN, F. K. (1968). *Molec. gen. Genet.*, **102**, 247–256.

DISCUSSION

Auerbach: Fifteen years ago Evelyn Witkin said that the results with liquid post-incubation, which you used, were too complicated so she has used transfer on membranes. Corran in my unit worked on mutations induced by an alkylating agent in *Bacillus subtilis*. He got the typical Witkin result when he transferred on membranes, using only solid medium. When he used liquid post-incubation he got a mutation frequency decline independent of whether he added broth or only required supplements (Corran, 1968). Something which is not simply MFD happens when liquid is used.

Clarke: Evelyn Witkin *did* use this method (Witkin, 1966*a*), and as far as I know, she has never used membranes: she always washes off solid medium (Witkin, 1958).

Bridges: When she described these Mfd⁺ or Mfd⁻ mutants in 1966/7 she was using a liquid medium.

Auerbach: When I told her about the experiments with *B. subtilis* which Corran did she said that with liquid one gets the most surprising results.

Maaløe: What was the temperature of the plates that you used, Dr. Clarke? We have seen striking decreases in survival levels when plating on pre-warmed rather than room temperature plates.

Clarke: I don't use *cold* plates.

Auerbach: When Corran (1968) post-incubated in liquid, he found the same mutation frequency decline whether or not he added broth to the incubation medium. What happens if you incubate in liquid containing broth?

Clarke: I haven't done that (but see Doudney and Haas, 1959). Basically my experiments repeat Evelyn Witkin's experiments (1966*a*). They apply to both kinds of mutations, they measure MFD for longer than an hour, and they involve plating on both broth-free and broth-rich media for survival and mutation.

Witkin (Comments added in proof): I have studied MFD in liquid and on solid media, and have found that there is no essential difference, although it may sometimes look as if there is. I have found that MFD occurs in nutrient broth, but not on nutrient agar, as Dr. Corran did. In our case, we found that protein synthesis occurred at a sub-maximal rate in nutrient broth, which accounted for the MFD we obtained, while on nutrient agar protein synthesis occurred at the maximal rate. The explanation is simply that in liquid post-treatments one cannot afford to dilute as much as is normally done in plating, and the cell densities we used in broth were too high for optimal synthesis. At equal cell number: medium volume ratios, nutrient broth and nutrient agar are equivalent. MFD will occur even on nutrient agar, if the plated population exceeds a certain number.

The main difference between demonstrating MFD in liquid and on plates lacking an amino acid pool is that the period during which MFD can occur on these plates, before active protein synthesis begins, is very much longer than the limited periods of time usually used for liquid treatments before plating on enriched medium. If the time period between u.v. irradiation and resumption of active protein synthesis is carefully equated, liquid and solid MFD-promoting treatments are entirely equivalent in my experience.

Magni: Was the molecular nature of the reversions checked in all these experiments? You said that one mutant was ochre and the other one was amber, Dr. Clarke. Are the revertants due to true back mutation of the mutant triplet or are they due to suppressor mutations? And what is the proportion of the two kinds of mutation?

Clarke: We do not yet have the answers to these questions.

Magni: This is very important because external or internal factors changing the mutation rate may affect diverse molecular events differently. For example in yeast some mutants show a "medium effect" in their reversion, and others do not.

Kimball: When you say reversions what kind of mutations are you talking about? Are you talking about base substitutions?

Magni: I say reversions because the revertant frequency is the parameter operationally checked. I posed my question because in yeast we have evidence that frame-shift mutations are insensitive to the medium effect and the same holds true for true back mutation of mis-senses and nonsenses. The only reversions which appear to vary in their frequency seem to be those due to suppressors (either mis-sense or nonsense suppressors).

Apirion: Yeast spores might be less likely to leak out molecules after treatment, and the difference in response between bacteria and yeast could perhaps be attributed to such effects.

Bridges: We haven't looked at MFD with the same strain as you, Dr. Clarke, although in your strain we had previously found that almost all the mutations are nonsense suppressors. We used WP2, which is a tryptophan-requiring strain containing an ochre triplet, to study mutation frequency decline. We distinguished between true mutations and ochre suppressor mutations on the basis of their ability to be lysed by nonsense T4 phages. Only the suppressor mutations showed mutation frequency decline. We didn't get any effect with the true mutations.

Clarke: One would predict that on the broth-enriched plates a high percentage of revertants would be amber and ochre suppressors and on the broth-free plates a relatively high percentage would be the true mutations.

Bridges: No, there are still about 90 per cent suppressor mutants although the proportion is of course lower than on broth-enriched plates (about 98 per cent).

Apirion: What do you mean by "true"?

Bridges: These are mutations which we presume to be true reversions on two counts. First, they don't contain any detectable nonsense suppressors, and secondly they grow as fast as the wild-type strain.

Dawson: The nature of the genetic change in the reversions should certainly be known, because the system could be summing a very large number of different genetic changes, and such a summation would make the experimental data almost uninterpretable.

Auerbach: I hoped that Professor Maaløe would be able to suggest how one could test whether the broth effect is due to the fact that a new transfer RNA has to be formed for suppressors.

Clarke: Evelyn Witkin (1956) did a reconstruction experiment in which she transduced a *try*$^+$ allele into a *try*$^-$ strain in an attempt to find out what is the delay, or what is the broth dependence, for expression of the *try*$^+$ gene suddenly introduced into *try*$^-$ cytoplasm. Donald MacPhee at Norwich has suggested transducing either a *try*$^+$ or an *su*$^+$ from a *try*$^+$ *su*$^+$ strain of *Salmonella* into a *try*$^-$ strain. When these *try*$^+$ and *try*$^-$ *su*$^+$ are plated on different media there might be a differential effect on the expression of true revertants as compared with suppressors.

Apirion: This difference could be trivial and attributed to the physiological events that have to follow introduction of a wild-type gene or a suppressor gene for a specific protein. In the first case messenger production and protein synthesis would follow soon after introduction of the wild-type DNA, while in the second case the suppressive agents (tRNA, ribosomes etc.) first have to be built up, and only then can they exercise their corrections, which are only partial.

Clarke: Part of the broth effect is due to the suppressors growing more slowly than the others on plates without broth.

Maaløe: What does "late log phase" really mean? The cell densities you mentioned seem to mean that the inoculum you use is so large that exponential growth is not established before the culture goes into the decline phase. What would happen if you made a 1:1000 dilution?

Clarke: With real log phase cells we see somcthing different. All Dr. Witkin's experiments were done with late lag phase cells and Ruth Hill (1968) has already given some evidence that this phase is a bit of a hotch-potch; we suspect 10 per cent of the cells are really in log phase and the rest are not, so maybe these are the wrong conditions to use.

Maaløe: What you have said indicates that the effect is stronger the more actively the cells grow, i.e. when, say, a new tRNA species could be synthesized in the shortest possible time. Suppressor type revertants may grow so slowly that unless they get an early start other effects might kill the cell on the way.

Auerbach: A recent paper dealt with the effects of u.v. on transcription and translation under different nutritional conditions (Absuba, Okajima and Honjo, 1968). Transcription was delayed or inhibited under both enriched and deficient conditions, whereas translation was only inhibited in the presence of amino acids.

Maaløe: Hans Bremer (1969) has good evidence that transcription stops at dimer lesions and that incomplete messenger RNA is formed. This effect would interfere with the transcription of a short chain like a transfer RNA only at high u.v. doses, unless several transfer RNA cistrons are grouped and require one continuous process of transcription.

Apirion: Is there MFD for forward mutations?

Clarke: As far as I know the only system in which forward mutations have been studied is that of streptomycin resistance. Witkin and Theil (1960) claim there is no MFD for these conditions whereas Shankel (1962) said there is MFD under some conditions.

Magni: We have done some work on forward mutations for canavanine resistance in yeast. The molecular basis for canavanine resistance is a block in an arginine permease, so it could be compared to a deficiency. If any effect of post-treatment conditions exists, it is very limited. I would prefer to say there is no effect.

Bridges: Dr. Witkin (1966*a*) suggested that mutation frequency decline may be due to the ease with which excision repair enzyme can get at the suppressor loci. She suggested that in the repressed state the suppressor loci were more susceptible to repair; dimers could be excised more easily under those conditions. She was in fact trying to interpret mutation frequency decline in terms of excision repair of the DNA, which I think is still the best working hypothesis.

Auerbach: Why should it act specifically on suppressor mutations?

Bridges: Because when one takes tryptophan away the *RC rel* locus switches off transfer RNA and ribosomal RNA synthesis, but it doesn't cut off messenger RNA synthesis. One therefore wouldn't expect it to affect true mutations at loci where there is messenger synthesis. One would expect it to affect only suppressor mutations at loci where either transfer or ribosomal RNA is synthesized. S. Igali and I have been looking at this in a recently isolated relaxed mutant of WP2. When one removes tryptophan, as one does in a mutation frequency decline experiment, RNA synthesis goes on at a much higher rate than in the normal (stringent) strain. Conversely, mutation frequency decline is slowed down in the relaxed strain to some extent. So there does seem to be an inverse connexion between the rate of transfer RNA and ribosomal RNA synthesis and the rate of mutation frequency decline, which would be consistent with Dr. Witkin's idea.

Maaløe: You may be right, but when a relaxed strain has accumulated RNA of the stable kind, during starvation, then, for some time after the readdition of the amino acid both induction and derepression of enzyme synthesis are greatly inhibited. This has nothing to do with mutations; the inhibition phenomenon is transient, and it certainly has nothing to do with the need for making a new species of tRNA. This mysterious effect has been studied in detail by G. Turnock (unpublished).

Bridges: It must presumably be something to do with the repressed state and what represses the transfer RNA locus. We don't know this of course.

Clarke: Is the relaxed mutant in fact slower in MFD? One would then suspect that an Mfd⁻ mutant might map in the *RC* locus. What I am really saying is that we should isolate Mfd mutants and map them.

Bridges: Mapping doesn't necessarily tell you anything.

Clarke: No, but if they all fall into the *uvrA* locus, that helps.

Bridges: It is quite possible that Mfd⁻ mutants could be relaxed mutants although one we have recently examined was not.

Witkin (*Comments added in proof*): I do not agree with Dr. Clarke that the final nature of MFD is still unclear. I think that the evidence overwhelmingly supports the interpretation that I offered several years ago: that MFD is a special case of excision repair, in which certain premutational u.v. photoproducts in suppressor genes (but not in most other genes) are excised only if protein synthesis is inhibited immediately after irradiation.

Several independent lines of evidence lead to this conclusion, but the single most telling fact is a property of Mfd⁻ mutants which is apparently not known to Dr. Clarke, although I have based my published argument on this evidence above any other (Witkin, 1966*a*, *b*). I refer to the fact that Mfd⁻ mutants excise pyrimidine dimers from their DNA at a drastically reduced rate compared to their Mfd⁺ parents. In this sense, Mfd⁻ mutants are excision-defective, although the defect is only in the extremely slow rate of excision, and not in any reduction of the amount of lethal damage ultimately excised, since Mfd⁻ strains are as u.v. resistant as their Mfd⁺ parents. Of 15 unrelated Mfd⁻ mutants isolated in my laboratory, all selected for poor MFD ability, four have been tested by Dr. J. Setlow, and have been found to excise pyrimidine dimers at an extremely slow rate. Although the other 11 have not had their excision rate measured directly, they all show the two properties that led me to suspect the first Mfd⁻ mutant of having an abnormally slow rate of excision: a greatly prolonged post-irradiation lag in cell division, and an extremely slow rate of decay of photoreversibility of u.v. killing. No Mfd⁻ mutant has ever been found which fails to exhibit these properties. It is therefore very probable that inefficient MFD and slow pyrimidine dimer excision are inseparable consequences of a single mutation.

Slow pyrimidine dimer excision and inefficient MFD are produced not only as simultaneous effects of a single mutation, but also as inseparable phenotypic effects of post-treatment with caffeine or acriflavine in Mfd⁺ strains. Caffeine retards MFD (Witkin, 1958) and inhibits pyrimidine dimer excision (Sideropoulos and Shankel, 1968). Acriflavine also inhibits both MFD and pyrimidine dimer excision (Witkin, 1961; and Setlow, 1964). Even the enhancement of u.v. mutation yields produced in Mfd⁺ strains by caffeine and acriflavine (Witkin, 1958, 1961) is duplicated by the Mfd⁻ mutation, which, as shown in my data (Witkin, 1966*a*) and as noted by Dr. Clarke, results in elevated yields of induced mutations, especially at extremely low u.v. doses. I have produced near-perfect phenocopies of an Mfd⁻ mutation by post-treatment of an Mfd⁺ strain with a concentration of caffeine chosen to duplicate the lag-lengthening effect of the Mfd⁻ mutation, i.e. to match excision rates. Thus, slowing the rate of pyrimidine dimer excision, either by an Mfd⁻ mutation or by post-treatment with an excision inhibitor, invariably results also in inhibition or retardation of MFD. The simplest explanation is that MFD is excision.

Another important property of Mfd⁻ mutants is that the photoreversibility of suppressor mutations is very inefficient in such strains, although their ability to split pyrimidine dimers enzymically is perfectly normal

(Witkin, 1966*a*). Acriflavine post-treatment of Mfd^+ strains causes similarly inefficient photoreversal of suppressor mutations, but does not affect photoreversal of u.v. killing (Witkin, 1963). In unpublished experiments, I have found caffeine post-treatment to have the same effect. The poor photoreversibility of suppressor mutations in Mfd^- strains strongly supports my earlier conclusion (Witkin, 1964) that MFD-susceptible lesions in suppressor genes are photoreversed indirectly, by a process requiring efficient dark repair after exposure to visible light, and that they are therefore not photoreversible pyrimidine dimers. They are either pyrimidine dimers which are not accessible to the photoreactivating enzyme, or they are not pyrimidine dimers at all. In either case, they are unique in their reparability. The possibility that MFD-susceptible lesions are structurally different from pyrimidine dimers is supported by J. Setlow's unpublished finding that these mutations have an induction action spectrum that differs from the induction action spectrum of suppressor mutations appearing on acriflavine-supplemented plates. The acriflavine-enhanced mutations are not MFD-susceptible, in the sense that they arise from photoproducts which are ordinarily excised whether protein synthesis is inhibited after u.v. or not.

MFD-susceptible u.v. photoproducts are thus not photoreversible directly by enzymic dimer-splitting, and they are excised only if protein synthesis is inhibited after irradiation. I think there can be little doubt that MFD is the excision of these exceptional photoproducts, but it is not yet clear why the inhibition of protein synthesis specifically improves their excisability. Since suppressor genes code elements of translation, such as transfer RNAs, it is likely that turning off protein synthesis triggers the repression of these genes. If the repressed form of a gene is more readily reparable than the active form, which I have suggested (Witkin, 1966*a*, *b*), the exceptional lesions, not excisable when the suppressor genes are active, may be rendered excisable as soon as these genes are repressed. We have recently found that enzymic dimer-splitting in suppressor genes, in excision-defective Hcr^- strains, is greatly increased under the same conditions that promote MFD in Hcr^+ strains, i.e. under conditions probably resulting in the repression of these genes.

Finally, I would like to comment on Dr. Clarke's conclusion that MFD can be explained, in large part, as a consequence of alteration of survival during the MFD-promoting treatment. Although under some conditions, and in some strains, changes in survival may occur during MFD treatments, and although these may influence the apparent magnitude of MFD when they do occur, I most emphatically reject the implication that MFD is largely an artifact of survival shifts. For nearly 15 years, I have studied MFD in dozens of different strains, at all stages of the culture cycle, and under a greater variety of conditions than I can easily count. There is no doubt whatsoever that "classical" MFD occurs under conditions resulting in no detectable change in overall survival (at survival levels low enough to detect such changes should they occur), as well as under some other conditions that cause slight increases or decreases in survival. In my experience, survival during MFD-promoting treatments, when it does not remain stable, tends to decrease slightly much more often than it increases, ruling out dose-reduction as a factor in MFD. A careful study of the extensive literature of MFD should make this clear (see, for example, Table I in Witkin, 1958). The trivial role, if any, of minor shifts in survival, when they occur, is most clearly demonstrated by the failure of u.v.-induced mutations to streptomycin resistance to exhibit MFD, when scored from the same chloramphenicol treatment tube which promotes typical MFD of suppressor mutations (Witkin and Theil, 1960). Any decline in mutation frequency explainable as due to survival changes should surely have affected both kinds of mutations.

REFERENCES

ABSUBA, J., OKAJIMA, S., and HONJO, J. (1968). *Int. J. Radiat. Biol.*, **14**, 517.
BREMER, H. (1969). *J. molec. Biol.*, in press.
CORRAN, J. (1968). *Molec. gen. Genet.*, **103**, 42.
DOUDNEY, C. O., and HAAS, F. L. (1959). *Proc. natn. Acad. Sci. U.S.A.*, **45**, 709–722.
HILL, R. F. (1968). *J. gen. Microbiol.*, **52**, 261–270.
SETLOW, R. B. (1964). *J. cell. comp. Physiol.*, **61**, suppl. 1, 51–68.
SHANKEL, D. M. (1962). *J. Bact.*, **84**, 410–415.
SIDEROPOULOS, A. S., and SHANKEL, D. M. (1968). *J. Bact.*, **96**, 198–204.
WITKIN, E. M. (1956). *Cold Spring Harb. Symp. quant. Biol.*, **21**, 123–140.
WITKIN, E. M. (1958). *X Int. Congr. Genet.*, Montreal, **1**, 280–299.
WITKIN, E. M. (1961). *J. cell. comp. Physiol.*, **58**, suppl. 1, 135–144.
WITKIN, E. M. (1963). *Proc. natn. Acad. Sci. U.S.A.*, **50**, 425–430.
WITKIN, E. M. (1964). *Mutation Res.*, **1**, 22–36.
WITKIN, E. M. (1966*a*). *Science*, **152**, 1345–1353.
WITKIN, E. M. (1966*b*). *Radiat. Res.*, suppl. 6, 30–53.
WITKIN, E. M., and THEIL, E. C. (1960). *Proc. natn. Acad. Sci. U.S.A.*, **46**, 226–231.

ENHANCEMENT AND DIMINUTION OF ULTRAVIOLET-LIGHT-INITIATED MUTAGENESIS BY POST-TREATMENT WITH CAFFEINE IN *ESCHERICHIA COLI* *

EVELYN M. WITKIN AND EGBERT L. FARQUHARSON

State University of New York Downstate Medical Center, Brooklyn, New York

MORE than ten years ago, caffeine was shown to increase the lethal and mutagenic effects of ultraviolet light (u.v.) when added to the post-irradiation plating medium in concentrations not in themselves toxic or mutagenic (Witkin, 1958). "Mutational synergism" was confirmed by Lieb (1961) and by Shankel (1962), and was interpreted, several years before the discovery of excision repair, as a consequence of inhibition by caffeine of enzymic dark repair of premutational lesions in DNA (Witkin, 1961; Lieb, 1961). This hypothesis gained support when Sauerbier (1964) found that caffeine inhibits "host cell reactivation" of irradiated bacteriophages and when caffeine was shown to inhibit the excision of pyrimidine dimers from DNA (Sideropoulos and Shankel, 1968). These observations, and especially the absence of both lethal and mutational synergism between caffeine and u.v. in an excision-defective (Hcr^-) strain (Clarke, 1967; Sideropoulos and Shankel, 1968), have led to the conclusion that caffeine enhances u.v. effects in Hcr^+ strains by decreasing the number of potentially lethal and potentially mutagenic u.v. photoproducts removed from the DNA by excision.

If caffeine modifies u.v. effects only by inhibiting excision repair, it should exert no influence on u.v. survival or u.v. mutagenesis in excision-defective strains. While both lethal and mutational synergism between caffeine and u.v. have been reported absent in strain WWP2 (Hcr^-), Clarke (1967) has noted that caffeine appears to depress the yield of u.v.-induced mutations to prototrophy slightly in this strain, and a similar effect can be discerned in the data of Sideropoulos and Shankel (1968) for induced streptomycin resistance. Clarke suggests that this may be either a weak antimutagenic effect of caffeine, or merely the consequence of a slight, undetected reduction in survival. If caffeine actually diminishes u.v.

* Weather conditions prevented Dr. Witkin from reaching London and presenting this paper, which therefore could not be discussed at the meeting.

mutagenesis in Hcr⁻ strains, it may do so by affecting the other dark-repair mechanism now known to promote survival in u.v.-irradiated *E. coli*, i.e. post-replication repair or genetic reconstruction, which operates independently of excision repair, and is also known to occur in Hcr⁺ strains (Rupp and Howard-Flanders, 1968; Howard-Flanders, 1968). This repair mechanism does not act on the pyrimidine dimers themselves, but on secondary u.v. lesions (i.e. daughter-strand gaps) induced by unexcised pyrimidine dimers in replicating DNA, and it probably involves some form of genetic recombination between sister strands (Howard-Flanders, 1968). It has been postulated (Witkin, 1968) that u.v.-induced mutations originate as errors in the recombinational repair of DNA gaps, a hypothesis strongly supported by the u.v. stability of *recA* strains, which lack recombination ability (Miura and Tomizawa, 1968; Witkin, 1969), and by the reduced u.v. mutability of *recC* strains, which have a low level of recombination ability (Witkin, 1969). Since strains having an *exr*⁻ mutation are also u.v.-stable (Witkin, 1967), the products of both the *recA*⁺ and the *exr*⁺ alleles must be necessary for u.v. mutagenesis to occur. The *recA* product is necessary for any recombinational repair (Howard-Flanders, 1968), and the *exr* product appears to be responsible for the inaccuracy of this repair (Witkin, 1968). Caffeine could diminish u.v. mutagenesis in Hcr⁻ strains by modifying the process of recombinational repair in a way that reduces its inaccuracy.

The slight diminution of u.v. mutagenesis observed by Clarke (1967), and evident also in the data of Sideropoulos and Shankel (1968), was produced by post-treatment with 0·05 per cent caffeine. This report describes experiments carried out to determine whether higher concentrations of caffeine might show a more pronounced antimutagenic effect in u.v.-irradiated Hcr⁻ strains, and if so, whether caffeine diminishes u.v. mutagenesis by modifying the process of recombinational repair.

MATERIAL AND METHODS

Bacterial strains

Strains WP2, a tryptophan-requiring derivative of *E. coli* B/r having normal excision ability, and WWP2 (Hcr⁻), an excision-defective derivative of WP2 isolated by Dr. R. Hill and kindly supplied by her, were used in most of the experiments.

The six Hcr⁺Exr⁺ strains used to obtain the data in Table I were: B/r, WP2, H/r30, H/r30-R, D13 and 26xA3. The five Hcr⁻Exr⁺ strains were: WWP2 (Hcr⁻), H_s30, H_s30-R, 26xA2 and 175A1. The three Hcr⁻Exr⁻

strains were: R29, 26x and R-53. All of these strains are Rec^+ and Fil^-, and all were derived originally from wild-type *E. coli* B.

The Hfr strain used as a donor to obtain the data in Table II was the K12 strain CS101, obtained from Dr. R. Alexander, which requires methionine and is streptomycin-sensitive. The F^- recipients were all streptomycin-resistant variants of the following *E. coli* B derivatives: WWP2 (Hcr^-), H_s30 and H_s30-R. All three of these strains are excision-defective and auxotrophic. H_s30 and H_s30-R require arginine, and were obtained from Dr. S. Kondo.

All strains used in this study had approximately the same sensitivity to caffeine, as determined by parallel streaking on gradient plates containing a range of caffeine concentrations from none to 1·0 per cent. For all strains, confluent growth stopped sharply at a level corresponding to about 0·3 per cent caffeine.

Preparation and irradiation of cultures

Cultures were grown and prepared for irradiation, and were exposed to u.v., as previously described (Witkin, 1963). Saline suspensions of cells in the late lag phase, having a titre of about 10^8 bacteria per millilitre, were irradiated.

Plating media

Difco Nutrient Agar, with 0·5 per cent NaCl added, was used as the plating medium, except in experiments involving mutations to prototrophy, in which case 5 per cent semi-enriched minimal (SEM) agar was used to score both survival and mutation frequency (Witkin, 1963). Caffeine, obtained from National Biochemicals Corporation, was added to the plating medium before autoclaving, in final concentrations of 0·1 per cent or 0·2 per cent.

Mutational systems

Mutations to streptomycin resistance were selected as previously described (Witkin and Theil, 1960), except that the plated population was allowed to increase tenfold before the antibiotic was added. This time varied from two to six hours. Colony counts were made after six days of incubation at 37°C.

Mutations to tryptophan independence were scored as previously described (Witkin, 1963), except that colony counts were made after four days. The longer incubation times were necessary because of relatively slow growth on media containing 0·2 per cent caffeine.

Genetic recombination

Crosses were made by mixing 1 ml of the recipient with 0·1 ml of the donor, using logarithmically growing cultures in Difco Nutrient Broth having a titre of about 2×10^8 bacteria/ml. The mixtures were incubated without agitation for 100 minutes at 37°, and were then agitated at high speed for one minute on a Vortex mixer to interrupt mating, and various dilutions were plated on the surface of minimal agar plates containing 100 μg streptomycin/ml and a small amount of Difco Nutrient Broth (1·25 per cent by liquid volume), with or without 0·2 per cent caffeine. This medium was selective for prototrophic recombinants, which were counted after three days.

RESULTS

Table I summarizes the effects of caffeine on u.v. survival and on the post-irradiation lag in cell division in three classes of strains differing in Hcr and Exr phenotypes. In all Hcr^+Exr^+ strains examined, the first post-irradiation cell division was greatly prolonged by caffeine, although the degree of enhancement of the lethal effect of u.v. was variable, and one strain (D13) showed no such enhancement. On the average, both the amount of enhancement of u.v. killing and the time added to the lag in cell division were significantly reduced in the Hcr^-Exr^+ strains, although in all of these strains caffeine still caused a significant increase in the division lag, and all but one (WWP2 (Hcr^-)) showed some enhancement of the lethal effect of u.v. None of the Hcr^-Exr^- strains examined showed any lethal synergism between caffeine and u.v., nor did any of these strains take longer to complete the first post-irradiation cell division in the presence of 0·1 per cent caffeine than in its absence. Thus, excision-defective strains (if they are Exr^+) may or may not show residual lethal synergism between 0·1 per cent caffeine and u.v., but all show residual prolongation by caffeine of the post-irradiation lag in cell division.

TABLE I

EFFECTS OF CAFFEINE IN THE POST-IRRADIATION PLATING MEDIUM ON SURVIVAL AND ON THE LAG IN CELL DIVISION

Phenotype	*No. of strains*	*Lethal enhancement index**		*Lag index†*	
		Range	*Average*	*Range (min)*	*Average*
Hcr^+Exr^+	6	0·04–1·0	0·35	120–210	170
Hcr^-Exr^+	5	0·6–1·5	0·85	36–59	46
Hcr^-Exr^-	3	1·0	1·0	0	0

* % survival on nutrient agar containing 0·1 % caffeine at u.v. dose resulting in 1 % survival on nutrient agar without caffeine.

† Minutes added to time required to double initial viable count, after irradiation with u.v. dose resulting in 10 % survival on nutrient agar, by adding caffeine (0·1 %) to plating medium (corrected for slight caffeine-induced extension of lag in unirradiated controls).

Fig. 1 shows the effects of 0·1 and 0·2 per cent caffeine, added to the post-irradiation plating medium, on the frequency of u.v.-induced mutations to streptomycin resistance in strain WWP2 (Hcr⁻). These concentrations of caffeine have no effect on colony-forming ability or on

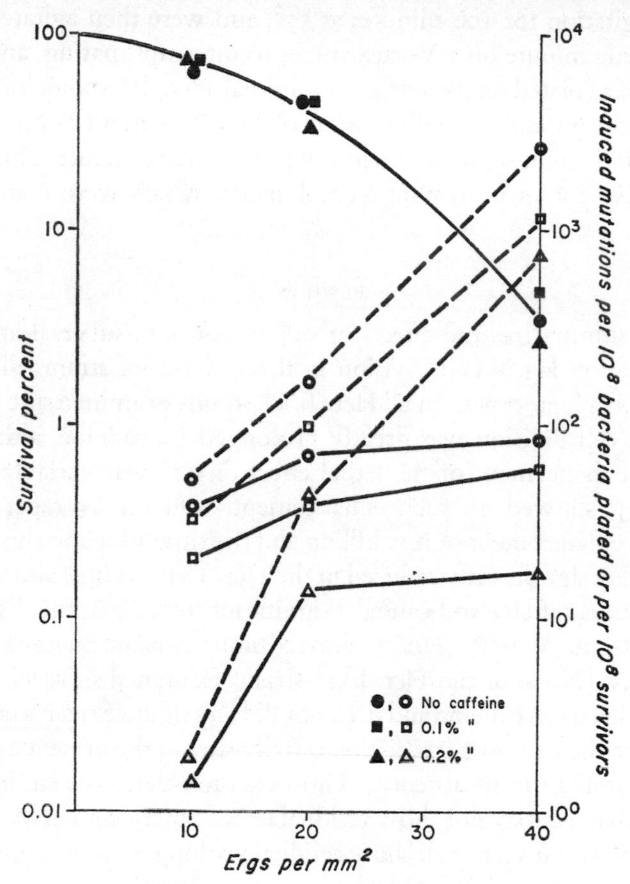

FIG. 1. Effect of caffeine on u.v. survival and on the yield of u.v.-induced mutations to streptomycin resistance in strain WWP2 (Hcr⁻).

Closed symbols: survival. Open symbols: mutations.

Open symbols, solid lines: induced mutations per 10^8 bacteria plated.

Open symbols, dashed lines: induced mutations per 10^8 survivors.

the spontaneous rate of mutation to streptomycin resistance in unirradiated WWP2 (Hcr⁻), although 0·1 per cent caffeine reduces the growth rate slightly, and 0·2 per cent caffeine does so more markedly. In the experiment shown in Fig. 1, neither concentration of caffeine affects u.v. survival, although in some experiments 0·1 per cent caffeine increases survival slightly, and in some experiments 0·2 per cent caffeine reduces survival slightly, but never to less than 60 per cent of the survival obtained without caffeine. The frequency of mutations to streptomycin resistance is significantly diminished by plating in medium containing 0·1 or 0·2 per cent caffeine, the magnitude of the effect increasing with caffeine concentration and decreasing as the u.v. dose is increased.

Fig. 2 shows results of a similar experiment to determine the effect of 0·1 and 0·2 per cent caffeine post-treatment on the yield of u.v.-induced

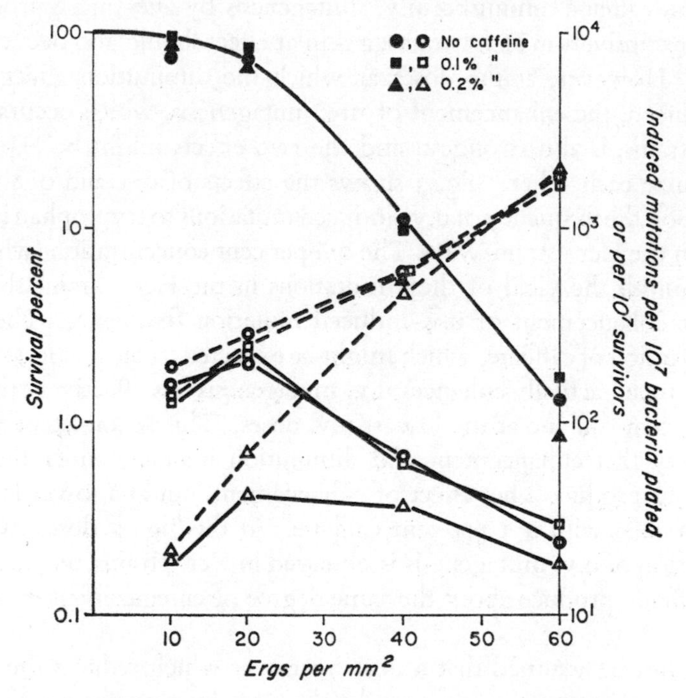

FIG. 2. Effect of caffeine on u.v. survival and on the yield of u.v.-induced mutations to tryptophan independence in strain WWP2 (Hcr^-).

Closed symbols: survival. Open symbols: mutations.
Open symbols, solid lines: induced mutations per 10^7 bacteria plated.
Open symbols, dashed lines: induced mutations per 10^7 survivors.

mutations to tryptophan independence in the same strain. The yield of these mutations, unlike that of mutations to streptomycin resistance, is not diminished by the presence of 0·1 per cent caffeine in the plating medium, but is greatly reduced at the lower doses by 0·2 per cent caffeine, which also reduces survival slightly but significantly. In all experiments, the effect on mutation yield is similar to that shown here, but 0·2 per cent caffeine does not always reduce u.v. survival perceptibly. As with streptomycin resistance, diminution of the mutation frequency by 0·2 per cent caffeine is most pronounced at the lowest u.v. dose used, and is almost entirely absent at the highest dose used.

Similar diminution of u.v. mutagenesis has been demonstrated for mutations to streptomycin resistance and to prototrophy in four other Hcr^- strains. In all cases, the results were similar to those obtained in strain WWP2 (Hcr^-).

Unless caffeine diminishes u.v. mutagenesis by affecting a process that occurs exclusively in Hcr^- strains, a similar effect should also occur in Hcr^+ strains. However, at the doses at which the diminution effect is most pronounced, the enhancement of u.v. mutagenesis, which occurs only in Hcr^+ strains, is also strongest, and the two effects might be expected to antagonize each other. Fig. 3 shows the effects of 0·1 and 0·2 per cent caffeine on the frequency of u.v.-induced mutations to tryptophan independence in the Hcr^+ strain WP2. The 0·1 per cent concentration (which does not diminish the yield of these mutations in the Hcr^- strain) shows the familiar enhancement of u.v.-induced mutation frequency. The higher concentration of caffeine, which might be expected to cause at least as much enhancement, actually enhances u.v. mutagenesis less effectively than does 0·1 per cent caffeine at the lowest u.v. doses. This is consistent with the possibility that enhancement and diminution interact, under these conditions, to produce a net effect of enhancement, but at a lower level than that obtained with 0·1 per cent caffeine. At the higher doses, where no diminution of u.v. mutagenesis is observed in Hcr^- strains, 0·1 and 0·2 per cent caffeine produce about the same degree of enhancement in the Hcr^+ strain.

It cannot be assumed that a post-treatment which reduces the yield of u.v.-induced mutations necessarily does so by preventing the actual occurrence of mutations (i.e. of stable changes in the DNA base sequence). Apparent antimutagenic effects can sometimes be caused by effects which prevent the detection of mutations in one of a great many possible ways without changing the number of mutations actually induced. The most obvious and most easily tested of these possibilities is differential sensitivity of the induced mutants to the agent in question. Ten u.v.-induced streptomycin resistant mutants and ten u.v.-induced Try^+ mutants were isolated, and each was mixed separately with the parent strain in reconstruction experiments designed to duplicate as closely as possible the conditions under which newly induced mutants are scored, with and without caffeine in the final plating medium. In all cases, the number of Str^r or Try$^+$ mutants recovered on 0·1 or 0·2 per cent caffeine-containing plates was the same as that recovered on plates without caffeine. Although there is no way to test the sensitivity of a newly induced mutant, as opposed to that of an established stock, it is certain, at least, that established stocks of Str^r and Try$^+$ mutants induced by u.v. in strain WWP2 (Hcr$^-$) are no more sensitive to caffeine than the parent strain.

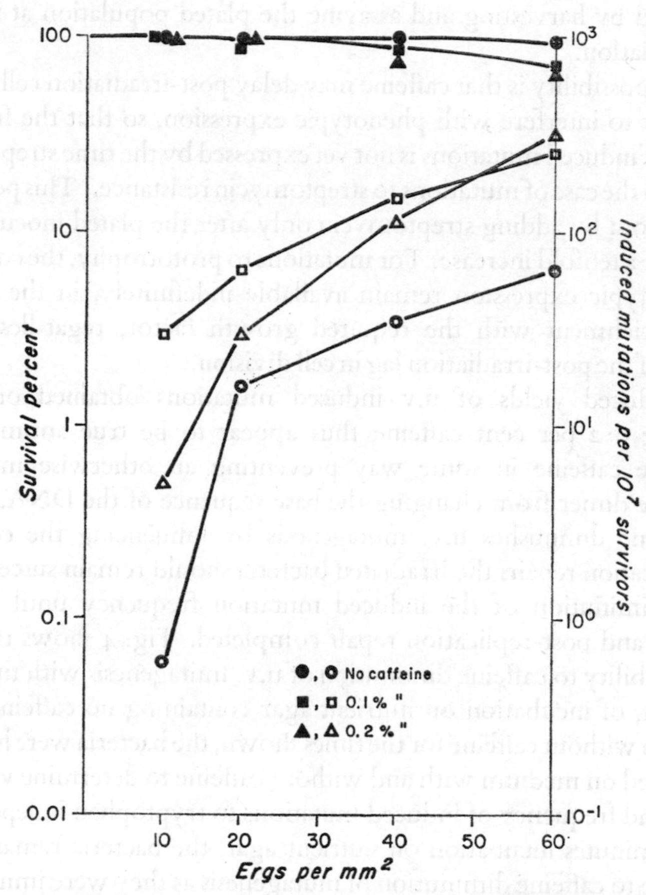

FIG. 3. Effect of caffeine on u.v. survival and on the yield of u.v.-induced mutations to tryptophan independence in strain WP2. Closed symbols: survival. Open symbols: mutations.

Another possible artifact is the assumption that survival of the irradiated bacteria in the presence of caffeine, as determined in the usual way by plating a high dilution, is necessarily the same on the heavily seeded plates used to score mutations. If caffeine kills irradiated bacteria on crowded plates more readily than those on uncrowded plates, the true frequency of induced mutations may be the same with and without caffeine. This was ruled out in two ways: first, by showing that the yield of induced Try^+ mutations is proportional to the dilution factor of the plated population over a wide range; second, by showing that the number of colony-forming units increases at the same rate in the presence of caffeine for dilute platings, as determined by periodic re-spreading, and for undiluted platings, as determined by harvesting and assaying the plated population at intervals after irradiation.

A third possibility is that caffeine may delay post-irradiation cell division sufficiently to interfere with phenotypic expression, so that the full complement of induced mutations is not yet expressed by the time streptomycin is added, in the case of mutations to streptomycin resistance. This possibility was ruled out by adding streptomycin only after the plated inoculum had undergone a tenfold increase. For mutations to prototrophy, the conditions for phenotypic expression remain available indefinitely, in the form of partial enrichment with the required growth factor, regardless of the duration of the post-irradiation lag in cell division.

The reduced yields of u.v.-induced mutations obtained on media containing 0·2 per cent caffeine thus appear to be true antimutagenic effects, the caffeine in some way preventing an otherwise mutagenic pyrimidine dimer from changing the base sequence of the DNA.

If caffeine diminishes u.v. mutagenesis by influencing the course of post-replication repair, the irradiated bacteria should remain susceptible to caffeine diminution of the induced mutation frequency until DNA is replicated and post-replication repair completed. Fig. 4 shows the decay of susceptibility to caffeine diminution of u.v. mutagenesis with time, after irradiation, of incubation on nutrient agar containing no caffeine. After incubation without caffeine for the times shown, the bacteria were harvested and replated on medium with and without caffeine to determine viable cell number and frequency of induced mutations to tryptophan independence. After 30 minutes incubation on nutrient agar, the bacteria remain just as susceptible to caffeine diminution of mutagenesis as they were immediately after irradiation. This susceptibility is lost to some extent between 30 and 60 minutes, but mainly between 60 and 90 minutes after irradiation, 90 minutes corresponding to the time required for doubling the initial viable count. These data indicate that caffeine acts to diminish u.v. mutagenesis relatively late in the first post-irradiation cell division, and are consistent with the possibility that it exerts this effect by influencing post-replication repair. Similar experiments with a thymine-requiring derivative of strain

WWP2 (Hcr⁻) showed that susceptibility to caffeine diminution of u.v. mutagenesis does not decay to any appreciable extent in 90 minutes if thymine is absent from the post-irradiation medium.

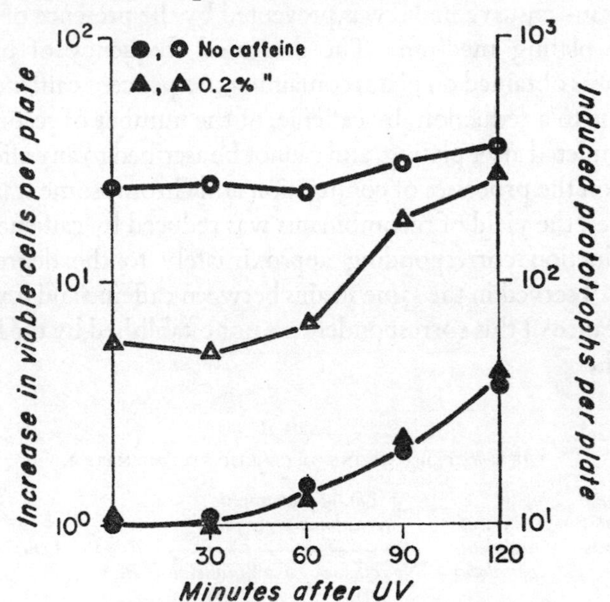

FIG. 4. Loss of susceptibility to caffeine diminution of u.v. mutagenesis with time of incubation on nutrient agar after u.v. irradiation.

Closed symbols: viable cells per plate.
Open symbols: induced prototrophs.

Irradiated bacteria were spread on surface of nutrient agar immediately after u.v. irradiation, incubated for the times indicated, and then harvested by washing the surface of two plates with sterile saline solution and assaying wash fluid on 5% SEM agar with and without caffeine to determine the viable count and number of prototrophic mutants.

If caffeine affects u.v. killing and u.v. mutagenesis in Hcr⁻ strains by influencing recombinational repair of single-strand gaps in DNA, caffeine might be expected to reduce the efficiency of genetic recombination when the same strains are used as recipients in conjugation. If reduced recombinational repair accounts for the enhancement of u.v. killing entirely, the extent to which genetic recombination is reduced should be more or less commensurate with the degree of lethal synergism observed. Table II shows results of an experiment in which the effect of 0·2 per cent caffeine on genetic recombination was explored. The three recipient strains are all streptomycin-resistant variants selected from excision-defective auxotrophs, all of which show decreased u.v. survival when 0·2 per cent caffeine is present in the plating medium. Conjugation and chromosome transfer were permitted to occur before plating, in the absence of caffeine. Mating was interrupted just before plating, and subsequent genetic activity of the streptomycin-sensitive males was prevented by the presence of streptomycin in the plating medium. The decreased frequency of prototrophic recombinants obtained on plates containing 0·2 per cent caffeine, therefore, must be due to a reduction, by caffeine, of the number of recombinational events completed after plating, and cannot be ascribed to any effects caffeine may have on the processes of conjugation and chromosome transfer. In all three crosses, the yield of recombinants was reduced by caffeine, the extent of the reduction corresponding approximately to the degree of lethal synergism observed in the same strains between caffeine and u.v., although the significance of this correspondence is not established by the limited data presented here.

TABLE II

THE EFFECT OF CAFFEINE ON GENETIC RECOMBINATION*

Strain from which Str^r recipient was derived	*No. of recipients per plate*†	*No. of prototrophic recombinants per plate*		*Ratio B/A*	*Lethal enhancement index*‡
		No caffeine (A)	*0·2% caffeine (B)*		
WWP2 (Hcr$^-$)	$2 \cdot 1 \times 10^7$	54, 50	30, 21	0·49	0·63
Hs30	$2 \cdot 2 \times 10^5$	106, 130	46, 46	0·39	0·51
Hs30-R	$1 \cdot 9 \times 10^5$	74, 83	24, 20	0·28	0·25

* For experimental procedure, see Material and Methods.

† Colony-forming ability of unirradiated recipient strains is not affected by presence of 0·2% caffeine.

‡ Survival of recipient strain on 5% SEM agar containing 0·2% caffeine after irradiation with a dose of u.v. which results in 1·0% survival on the same medium without caffeine (average of three determinations).

DISCUSSION

In the excision-defective strains, caffeine has residual effects indicating that its ability to influence u.v. survival and u.v. mutagenesis is not entirely due to its inhibitory effect on pyrimidine dimer excision. Harm (1967) has suggested that the residual lethal synergism between acriflavine and u.v. observed in Hcr$^-$ strains may reflect residual excision activity in such strains, and the same argument might be used for caffeine. In that case, however, both acriflavine and caffeine should cause some residual enhancement of u.v. mutagenesis in Hcr$^-$ strains, and neither agent does so. Acriflavine post-treatment fails to increase the yield of u.v.-induced mutations in Hcr$^-$ strains (Vechet, 1968; Witkin, unpublished observations), and caffeine post-treatment actually diminishes it, as shown above. While mutational synergism between these agents and u.v. can thus be fully accounted for by their inhibitory effect on the excision of pyrimidine dimers, lethal synergism persists in most Hcr$^-$ strains, and must, at least in part, have some other basis. It is likely that the residual enhancement of u.v. killing in Hcr$^-$ strains is due to effects of these agents on another survival-promoting mechanism which operates independently of excision repair, and which occurs in both Hcr$^+$ and Hcr$^-$ strains. Howard-Flanders (1968) has recently summarized the evidence indicating that survival in *E. coli* after u.v. irradiation is determined by the independent operation of two repair processes: excision repair (which removes pyrimidine dimers from the DNA before replication), and recombinational repair (which mends the daughter-strand discontinuities induced by unexcised pyrimidine dimers in replicating DNA).

While there is no proof that caffeine exerts its residual effects in excision-defective strains (i.e. slight enhancement of u.v. killing, moderate prolongation of division lag, marked diminution of u.v. mutagenesis) by influencing the process of post-replication repair after DNA replication, there are several independent indications that this may be so. One is the ability of caffeine to reduce genetic recombination after conjugation in Hcr$^-$ recipients to about the same extent that it enhances the lethal effect of u.v. in the same strains. Another is the late time of action of caffeine in the post-irradiation lag period, as indicated by the slow decay of susceptibility to caffeine diminution of u.v. mutagenesis, and by the failure of such decay to occur when DNA synthesis after u.v. is prevented. Perhaps the most significant indication is the absence of any residual effect of caffeine in Hcr$^-$Exr$^-$ strains. To appreciate why this observation implicates post-replication recombinational repair in caffeine diminution of u.v. mutagenesis, it is necessary to review some characteristics of *exr*$^-$ strains.

A mutation at the *exr* locus (which may be identical with the *lex* locus of K12) reduces both recombination ability and u.v. resistance to about 30 to 50 per cent of normal, increases both spontaneous and u.v.-induced DNA breakdown (Witkin, 1967 and unpublished observations), and increases sensitivity to X-rays (Mattern, Zwenk and Rörsch, 1966) and to nitrosoguanidine (Witkin, 1967), all of which suggest that this locus should be considered a fourth member of the *rec* group of loci, the products of which are required for normal recombination ability. In addition to these characteristics, *exr*$^-$ strains are u.v.-stable (Witkin, 1967), and show greatly reduced mutability towards γ-rays and thymine starvation, indicating that there is a common pathway for mutagenesis initiated by u.v., ionizing radiations and thymine starvation, in which single-strand gaps in DNA are repaired inaccurately when the *exr*$^+$ product is present (Bridges, Law and Munson, 1968). It has been postulated (Witkin, 1968) that *exr* mutations alter the recombination process in a way that reduces its efficiency relatively little, but entirely eliminates an error-prone step that makes the recombinational process itself mutagenic in Exr$^+$ strains. The absence of any demonstrable effect of caffeine post-treatment in irradiated Hcr$^-$Exr$^-$ strains is compatible with the possibility that caffeine exerts its effects in Hcr$^-$Exr$^+$ strains by inhibiting the product of the *exr*$^+$ allele, thus producing a weak phenocopy of an Exr$^-$ mutant.

Since caffeine is known to inhibit exonucleases in *E. coli* (Roulland-Dussoix, 1967), it is possible to account for its diminution of u.v. mutagenesis in another way. It has been proposed (Witkin, 1968) that the *exr*$^+$ allele codes a modifying enzyme which arrests DNA degradation, thus permitting recombination to occur, but in doing so changes the pairing specificity of the modified base. Caffeine may prevent the *exr*$^+$ gene product from participating in the recombinational repair process not by inhibiting it directly, but by preventing the exonuclease degradation of the DNA which may be a prerequisite for the occurrence of the mutagenic modification. In other words, caffeine may stabilize the DNA against degradation so that a recombinational event can be initiated without requiring a stabilizing (but mutagenic) modification of an end base by the *exr*$^+$ gene product. Puglisi (1968) has noted that many antimutagenic compounds inhibit DNA degradation, and has suggested that a certain amount of breakdown is a prerequisite for spontaneous mutation. Caffeine has also been reported to decrease spontaneous mutability under certain conditions (Grigg and Stuckey, 1966).

It is clear that caffeine enhances u.v. mutagenesis in Hcr$^+$ strains by inhibiting excision repair; it seems possible that it exerts the opposite effect in Hcr$^-$ strains by reducing the inaccuracy of recombinational repair. If this interpretation is correct, caffeine post-treatment should diminish γ-ray and thymine-less mutagenesis, and should extend the time required to restore the continuity of the daughter strands of DNA produced after u.v. irradiation of excision-defective strains.

SUMMARY

Caffeine, which enhances u.v. mutagenesis in Hcr$^+$ strains of *E. coli* when added to the post-irradiation plating medium, has an opposite effect in excision-defective (Hcr$^-$) strains, causing a significant diminution of u.v. mutagenesis in such strains. Caffeine post-treatment causes slight residual enhancement of u.v. killing in most Hcr^-Exr^+ strains, as well as residual prolongation of the post-irradiation lag, although both of these effects are more extreme in Hcr^+ strains. Caffeine post-treatment has no demonstrable effects in Hcr^-Exr^- strains.

It is concluded that mutational synergism between u.v. and caffeine, as observed in Hcr^+ strains, is entirely due to inhibition by caffeine of excision repair, while enhancement of u.v. killing and prolongation of the post-irradiation lag in such strains are only partly due to inhibition of excision repair. It is suggested that the effects of caffeine in Hcr^- strains may be due to its influence on recombinational repair of DNA gaps, such that the efficiency of this repair mechanism is slightly reduced (resulting in residual enhancement of u.v. killing and prolongation of the lag) while its accuracy is greatly increased (resulting in diminution of u.v. mutagenesis). It is suggested also that caffeine may produce these effects by inhibiting the activity of the *exr*$^+$ gene product.

Acknowledgement

This research was supported by grant No. 5 R01 A1 01240 from the National Institute of Allergy and Infectious Diseases.

REFERENCES

BRIDGES, B. A., LAW, J., and MUNSON, R. J. (1968). *Molec. gen. Genet.*, **103**, 266–273.
CLARKE, C. H. (1967). *Molec. gen. Genet.*, **99**, 97–108.
GRIGG, G. W., and STUCKEY, J. (1966). *Genetics, Princeton*, **53**, 823–834.
HARM, W. (1967). *Mutation Res.*, **4**, 93–110.
HOWARD-FLANDERS, P. (1968). *A. Rev. Biochem.*, **37**, 175–200.
LIEB, M. (1961). *Z. VererbLehre*, **92**, 416–429.
MATTERN, I. E., ZWENK, H., and RÖRSCH, A. (1966). *Mutation Res.*, **3**, 374–380.
MIURA, A., and TOMIZAWA, J. (1968). *Molec. gen. Genet.*, **103**, 1–10.
PUGLISI, P. P. (1968). *Molec. gen. Genet.*, **103**, 248–252.
ROULLAND-DUSSOIX, D. (1967). *Mutation Res.*, **4**, 241–252.
RUPP, W. D., and HOWARD-FLANDERS, P. (1968). *J. molec. Biol.*, **31**, 291–304.
SAUERBIER, W. (1964). *Biochem. biophys. Res. Commun.*, **14**, 340–346.
SHANKEL, D. M. (1962). *J. Bact.*, **84**, 410–415.
SIDEROPOULOS, A. S., and SHANKEL, D. M. (1968). *J. Bact.*, **96**, 198–204.
VECHET, B. (1968). *Folia microbiol., Praha*, **13**, 379–390.
WITKIN, E. M. (1958). *X Int. Congr. Genet.*, Montreal, **1**, 280–299.
WITKIN, E. M. (1961). *J. cell. comp. Physiol.*, **58**, Suppl. 1, 135–144.
WITKIN, E. M. (1963). *Proc. natn. Acad. Sci. U.S.A.*, **50**, 425–430.
WITKIN, E. M. (1967). *Brookhaven Symp. Biol.*, **20**, 17–55.
WITKIN, E. M. (1968). *XII Int. Congr. Genet.*, Tokyo, **3**.
WITKIN, E. M. (1969). *Mutation Res.*, **8**, 9–14.
WITKIN, E. M., and THEIL, E. C. (1960). *Proc. natn. Acad. Sci. U.S.A.*, **46**, 226–231.

ALLELE-SPECIFIC RESPONSES TO FACTORS THAT MODIFY U.V. MUTAGENESIS

BRIAN J. KILBEY

MRC Mutagenesis Research Unit, Institute of Animal Genetics, University of Edinburgh

We may define mutagen specificity as either the differential response of two or more genes to a particular mutagen, or the array of responses of a single gene to a number of different mutagens. The phenomenon may be illustrated by reference to some of the early results of Demerec (1953) with *Escherichia coli*. Table I shows the response of a diauxotrophic strain, leucine-less and phenylalanine-less, to three mutagens: manganous chloride, u.v., and β-propiolactone. The two mutants reverted at very different frequencies with each of the mutagens. Manganous chloride and β-propiolactone produced more leucine reversions than phenylalanine reversions; u.v. produced more phenylalanine reversions than leucine reversions. Comparison of the mutagens in terms of their effectiveness in producing a given mutational change is of limited value since nothing is known of the dose actually reaching the genes.

TABLE I

FREQUENCIES OF REVERSE BIOCHEMICAL MUTATIONS IN A DOUBLY MUTANT STRAIN OF *Escherichia coli* AFTER TREATMENT WITH THREE DIFFERENT MUTAGENS (from Demerec, 1953)

	Mutations per 10^8 *induced by*		
	$MnCl_2$	*U.v.*	*Lactone*
Leucine-less	594	57	28
Phenylalanine-less	11	100	3·3

From the example cited one might form the impression that the response of a particular mutant to a particular mutagen is invariable and depends solely on the mutant and mutagen concerned. This idea has been adopted as the basis for using specific reversibility tests in the analysis of the molecular basis of mutation. From their response to a number of selected mutagens, a high proportion of the *rII* mutant sites in bacteriophage T4 have been unambiguously assigned base pairs (Champe and Benzer, 1962). In cellular organisms, unambiguous assignment is more difficult but it is still often possible to place a mutant in a particular class of genetic alteration, e.g. frame shift or base substitution (de Serres, 1964). This difficulty arises, without doubt, from the modification of the primary mutational response by other cellular events. It has become very clear during the last few years that the response of a particular mutant to a mutagen can be drastically modified by ancillary treatments. Most of this work has been done with bacteria, using u.v., but examples are also on record for chemically induced mutations (Corran, 1968). The molecular basis for these phenomena has still to be elucidated in most cases. Repair activity is often assumed to be responsible, but it must be admitted that independent evidence for the involvement of repair is often lacking. However, whatever cellular events are responsible, the fact remains that the response of a gene to a mutagen is not fixed but is subject to a number of modifying factors.

Several cases are now known in which modifying factors are specific for a particular mutation. This is of great importance for the practical aspects of mutagen specificity, since it opens up the way for a certain amount of directed mutation treatment. The purpose of this contribution is to present an instance in which the response of two mutants to a particular mutagen can be modified by factors which are mutant specific. By means of these factors, which are themselves non-mutagenic, the specificity of the mutagen, u.v., can be considerably increased.

EXPERIMENTAL MATERIAL

Since the same mutational system forms the basis both for this discussion and the next paper, by Professor Auerbach, I shall describe it in detail here. Most of the experiments have been performed on the strain of *Neurospora* first isolated by Kølmark and given the isolation number K3/17. This strain carries the two mutants *ad-3A* 38701, *inos* 37401 which determine requirements for adenine and inositol, respectively. In addition, several colonial mutations have been accumulated which produce a small compact colony in the absence of those growth-limiting substances that are usually obligatory for the colonial growth of wild-type *Neurospora*. The details of the culture conditions have been described elsewhere (Kølmark and Kilbey, 1962). In addition to the K3/17 strain, another, synthesized independently by M. Allison, has also been used. It carries the same two auxotrophic mutations together with the mutation *cot* (colonial temperature-sensitive). At temperatures below 29°C this strain grows as freely as wild-type. Above 29° growth is restricted and small colonies are formed.

Conidia were harvested, washed and suspended in water for mutagenic treatment. The cell density normally used was 1 to 3×10^7 cells/ml. The suspension was either irradiated (for details see Kilbey, 1967) or treated with a chemical. In the latter case, treatment was terminated by centrifugation and washing, or by membrane filtration and washing.

By using suitable media, both types of reversion can be scored in the same experiment; adenine reversions may be selected on minimal medium plus inositol, inositol reversions on minimal medium plus adenine. Survival is scored on minimal medium plus both supplements. The plates were incubated for seven to ten days at 30° before scoring.

Mutagen specificity is conveniently expressed as the ratio of inositol to adenine reversions (i/a). Since the survival estimate at each dose is the same for both types of revertant, it can be ignored and the ratio can be calculated from the absolute numbers of revertants scored.

EFFECTS OF DOSE AND GENETIC BACKGROUND

In a paper dealing with the effects of several mutagens in K3/17, Westergaard (1957) reported that u.v. induces twice as many inositol as adenine reversions. The data in Fig. 1 show that this is only partly true. When

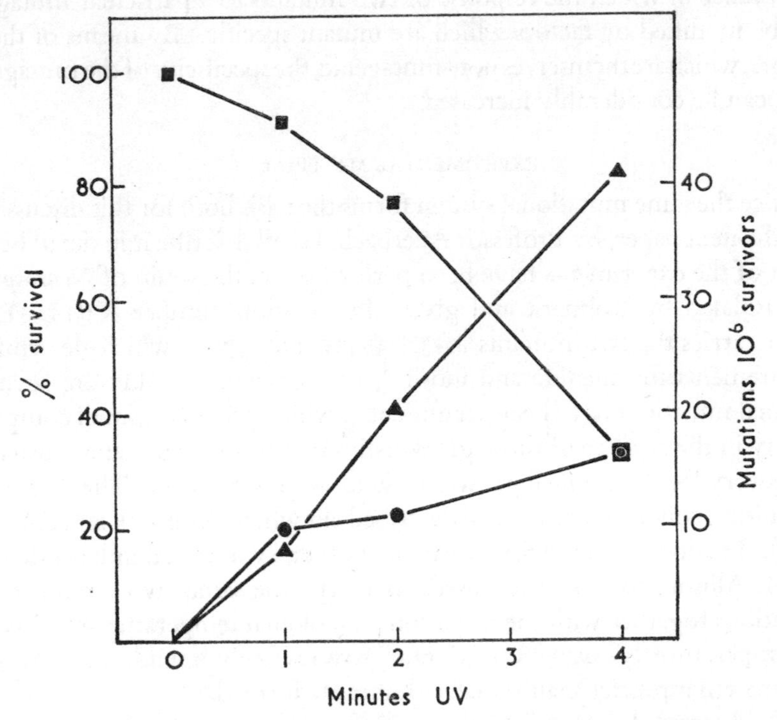

FIG. 1. The response of K3/17 to u.v. Squares: survival; triangles: inositol reversions; circles: adenine reversions.

K3/17 is treated with a series of u.v. doses the i/a ratio changes with dose. At low doses it is one or less than one, as in the experiment shown here. At higher doses it increases to between two and three. The increase in the i/a ratio with dose leads to divergence of the two dose-effect curves. This has also been observed by other workers with this strain (Kølmark, 1953; Auerbach and Ramsay, 1968).

Several explanations may be put forward to account for the increase in i/a ratio with dose. One is that inositol revertants might be selected because they are more resistant to u.v. than adenine revertants. Auerbach and

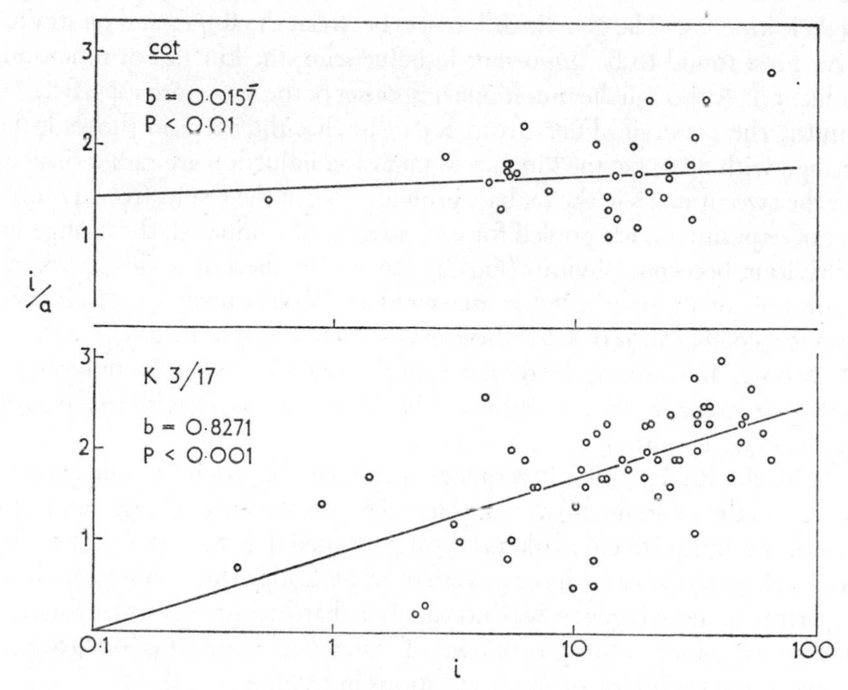

FIG. 2. The effect of genetic background on the variation of i/a with dose of u.v. measured in terms of inositol reversions (i). b: regression coefficient.

Ramsay (1968) ruled out this possibility by using mixtures of revertant and non-revertant cells in reconstruction experiments. If anything, inositol revertants proved to be slightly less resistant to u.v. than adenine revertants. Second, inositol revertants might be more sensitive than adenine revertants to the suppressive effects of the background of non-revertant cells on the mutation plates. The effect should be most noticeable at low doses and should decrease as more of the background cells are killed. This was excluded by plating u.v.-treated suspensions at different dilutions. Dilution did not alter the frequency of revertants scored. The final possibility considered by Auerbach and Ramsay was that, in the multinucleate conidia of strain K3/17, inositol reversions were less dominant than adenine reversions. This effect should be more acute at low than at high u.v. doses, when a reduction in the number of viable nuclei per conidium should occur. In fact, most of the reversions tested from low doses were heterokaryotic. Trivial explanations of this type do not appear to be responsible for the divergent mutation induction curves.

Divergence of the mutation induction curves, and the consequent rise in i/a ratio with dose, is an expression of the fact that the two curves differ in their kinetics. The genetic differences between the K3/17 and *cot* strains have been found to be important in influencing the kinetics of mutation induction. Although the mutational responses of the strains are superficially similar, the *cot* strain differs from K3/17 in that the i/a ratio shows little change with dose, i.e. the kinetics of mutation induction are rather similar for the two mutants in the *cot* background. When the results from a number of experiments are pooled for each strain and compared, the change in behaviour becomes obvious (Fig. 2). In K3/17 there is a clear positive regression of i/a upon dose as measured in inositol reversions. The regression coefficient is 0·8271, the standard error being 0·1504 ($t_{51}=5\cdot501$, $P<0\cdot001$). In contrast, the results from the *cot* derivative show only slight positive regression, the regression coefficient being 0·0157 with S.E. 0·0055 ($t_{28}=2\cdot833$, $P<0\cdot01$).

With the K3/17 strain, divergence was invariably found in our experiments; in the *cot* strain, however, some experiments showed slight convergence. Malling and co-workers (1959) presented data for the K3/17 strain which are at variance with the pattern of response described above. In their experiments convergence was noted. It is hard to give reasons for this discrepancy since nothing is known of differences in genetic background between the strains, or of slight variations in technique.

These results show clearly that the response of the two mutants is not fixed but can be modified, albeit only slightly, by altering either the u.v. dose or the genetic environment in which they are treated.

THE EFFECT OF TEMPERATURE

I would now like to turn to the effect of temperature on the response of the adenine and inositol mutants to u.v. In 1962, while working with the K3/17 strain, I discovered that the temperature of the conidial suspension during irradiation played an important role in determining the u.v.

response of the two mutants (Kilbey, 1963). Fig. 3 shows the result of a typical experiment which demonstrates this effect. At 30°C the response of the strain is as described earlier. The i/a ratios gradually increase with dose from 0·82 to 2·5. At 2° the ratios still increase with dose but they are all approximately half the corresponding values at 30°. This happens because, at the lower temperature, the frequency of adenine reversions is

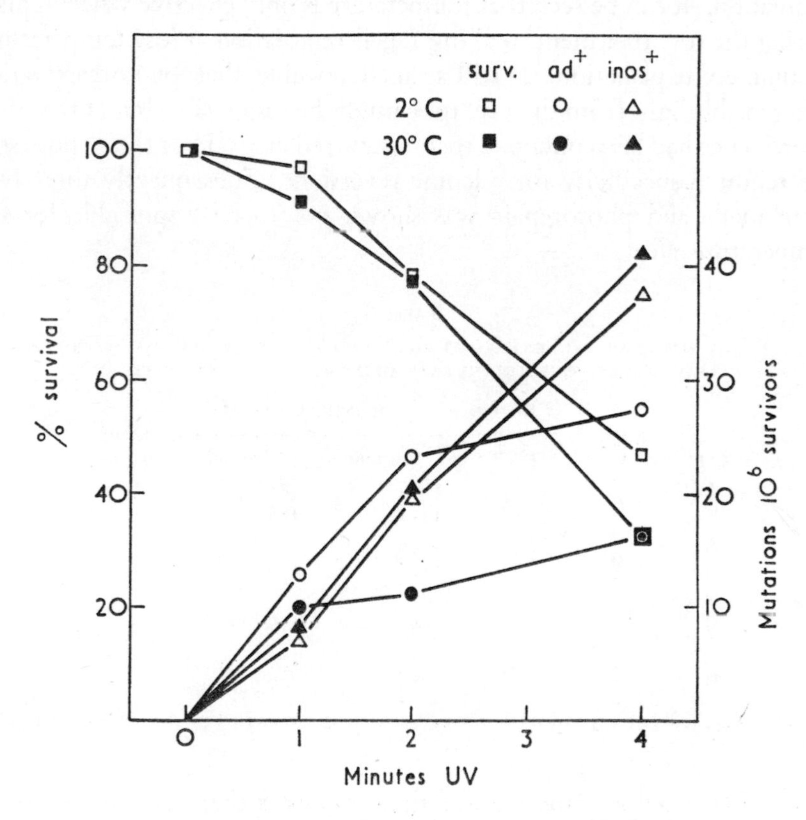

FIG. 3. The effect of temperature on the response of K3/17 to u.v.

increased specifically. There is little or no effect on either survival or inositol reversions. Reconstruction experiments showed that the effect could not be attributed to a greater u.v. sensitivity of adenine reversions at high temperatures.

It seems improbable that the temperature difference used in these experiments would greatly influence the photochemical events concerned in the induction of adenine reversions. In the first place the temperature coefficients for photochemical reactions are small, and secondly the coefficient would have to be negative. The more probable interpretation seemed to be that, at low temperature, slowing of repair brought about an increase in adenine revertants. If this were so, post-irradiation temperature differences might also be expected to modify mutation frequency. Table II shows the results of one of the experiments in which this was tested. U.v. exposures were given at high intensity for a short time to minimize repair during irradiation. It can be seen that temperature is only effective when applied during the u.v. treatment, making repair retardation at low temperatures an unlikely explanation. It still seemed possible that photoreactivation using visible light from the u.v. tube might be responsible but at that time no evidence had been obtained that it occurred in K3/17 or that it possessed the required specificity for adenine reversions. Subsequently these tests were made and photorepair was shown not to be responsible for the temperature effect.

TABLE II

THE EFFECT OF INCUBATION AT HIGH AND LOW TEMPERATURES AFTER ULTRAVIOLET EXPOSURE ON THE RATIO OF INOSITOL TO ADENINE REVERSIONS

Expt	*U.v. temp.* (°C)	*Post-u.v. temp.* (°C)	*Reversions counted* *Adenine*		*Inositol*		*Inositol/ adenine*
A	0	0	109	120·5	184	186	1·5
B			132		188		
A	0	25	109	125·5	193	201·5	1·6
B			142		210		
A	25	0	78		—		3·4
B			70		236		
A	25	25	92	85	363	286·5	3·4
B			78		210		

A and B are replicate experiments. Ultraviolet dose: 30 seconds; period of post-treatment: 30 minutes.

An alternative explanation for these results is that u.v., at the lower temperature, damages a cellular process, e.g. repair, which only becomes involved in the mutational pathway at a later stage. This is difficult to test at present.

For another mutagen it was clear that temperature acted at a later stage. Auerbach and Ramsay (1967) showed that temperature is important in determining the mutagen specificity of nitrous acid in K3/17. Here the temperature at which plated conidia were incubated was important. This post-treatment effect has two basic similarities to the u.v. experiments in that (*a*) the main effect was on the adenine reversions, (*b*) the higher temperature (32°) produced fewer reversions than the lower one (25°).

PHOTOREPAIR

Photorepair is the only repair system in fungi for which definite biochemical evidence exists at present. The enzyme has been extracted from *Neurospora* conidia and can reactivate u.v.-inactivated transforming principle *in vitro* (Terry, Kilbey and Branch-Howe, 1967). In earlier experiments Kilbey and de Serres (1967) were able to show that photorepair is apparently unspecific: all classes of u.v.-induced *ad*-3*B* mutants were diminished when photorepair was allowed to function. In view of this it was all the more surprising to find that in both the K3/17 and the *cot* strains, photorepair is specific for adenine reversions (Kilbey, 1967). Fig. 4 shows

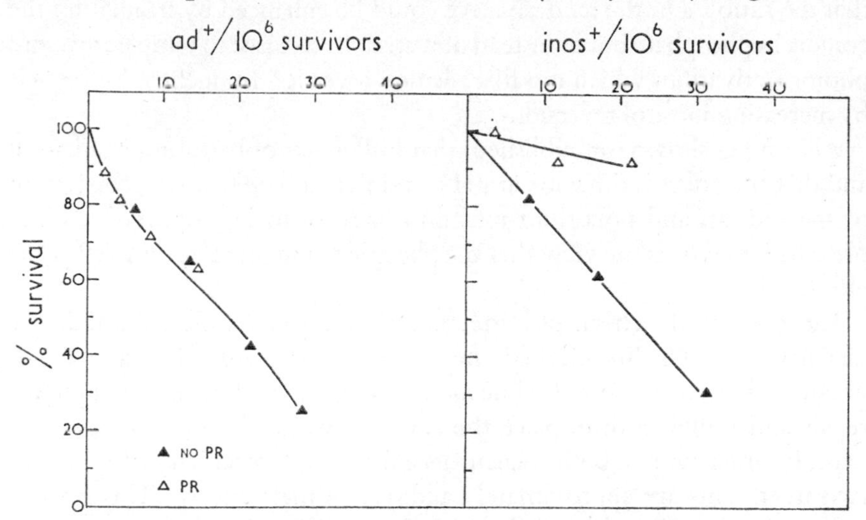

FIG. 4. Differential photoreactivation of mutation and survival in the *cot* derivative. PR: photorepair.

the two types of mutation plotted versus survival, with and without photorepair. The same dose reduction factor applies to both adenine reversions and survival; thus the points fall on one curve whether they come from photoreactivated samples or not. Inositol reversions are only slightly responsive to photorepair and, when plotted against survival, points from the light and dark samples fall on different curves.

The behaviour of the inositol mutant appears to be unique. Two other *inos* alleles were tested together with two *ad*-3*A* alleles and ten *ad*-3*B* alleles. All of these were photoreactivated to the same extent as *ad*-3*A* 38701 although their u.v. responses differed considerably.

The finding that photorepair operated specifically upon adenine reversions in this system led to a re-examination of the possibility that photoreactivation, caused by the visible light emitted by our u.v. source, was responsible for the temperature effect. Accordingly attempts were made to use the u.v. lamp as a photoreactivating source after first filtering the radiation through window glass to remove u.v. No photoreactivation was observed and it seems, therefore, that this cannot be the explanation for the temperature effect.

BUFFER DURING IRRADIATION

During experiments in which the pH dependence of a number of mutagenic treatments was being studied, Dr Allison in our laboratory found that the ratio i/a in the *cot* derivative could be enhanced by irradiating the conidia in phosphate buffer instead of water. In contrast to temperature and photoreactivation, which modify adenine reversion frequency, buffer acts by increasing inositol reversions.

Allison has shown (unpublished) that buffer acts only during irradiation and that the effect is the same at pH 6 and pH 7. The relative proportions of the sodium and potassium ions do not seem to be important and, at present, he favours the view that the phosphate ions are responsible for the effect.

Fig. 5 shows the effects of buffer and photorepair on the i/a ratio in the *cot* derivative. The line marked "no specificity" corresponds to an i/a ratio of unity; departures from this line indicate specificity. It is clear that photorepair and buffer both displace the curves towards the inositol revertant axis. In other words, both treatments enhance the specificity of u.v. The two treatments are approximately additive in their effects. This is to be expected since one enhances the i/a ratio by acting on adenine reversions and the other does it by enhancing inositol reversions.

Irradiation in buffer also modifies the kinetics of mutant induction in the *cot* strain. As shown earlier, treatment in water produces little or no divergence in this strain. In buffer there is a slight convergence. Evidence for this is apparent from the bends in the buffer curves in Fig. 5 and a more precise estimate of the effect can be obtained if data from several experiments are combined and plotted in the same way as in Fig. 2. A slight negative regression is obtained.

DISCUSSION

The effects of dose, genetic background, temperature, photorepair and buffer demonstrate the ability of experimental conditions to modify the mutational specificity of a particular mutagen in a particular mutational

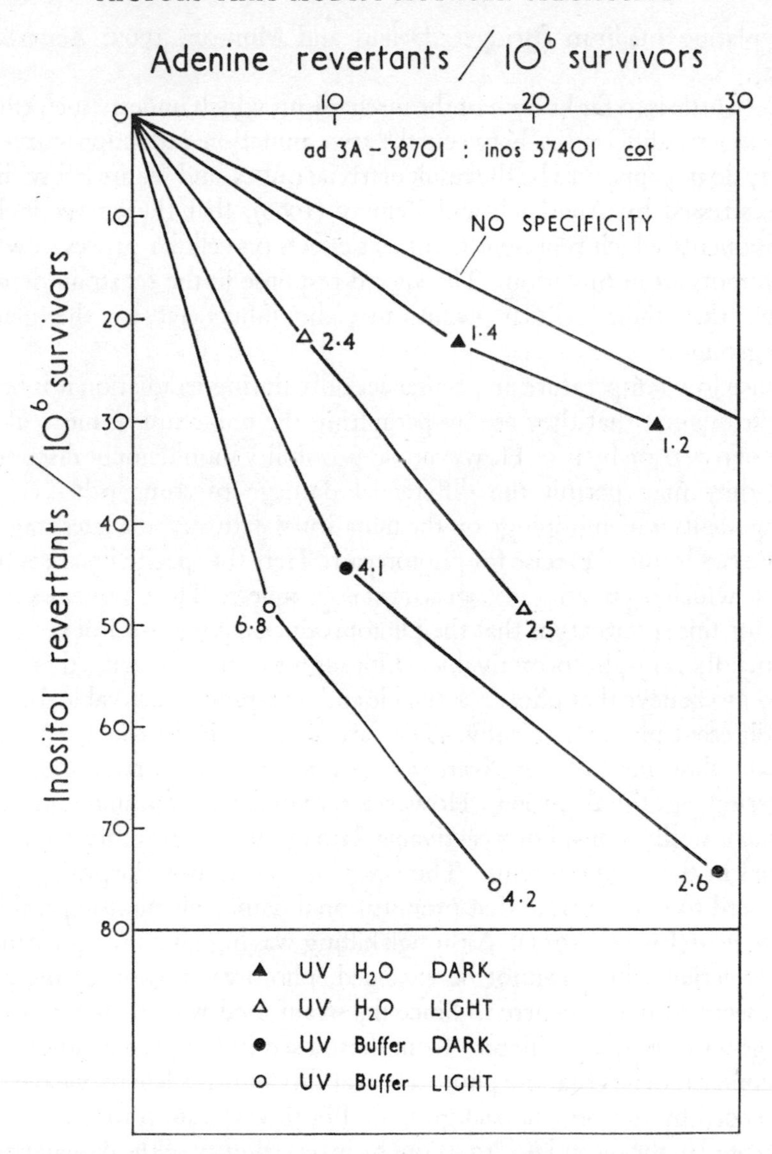

FIG. 5. The effects of photoreactivation and irradiation in buffer on the mutagen specificity of u.v. in the *cot* derivative.

system. Other instances have been recorded in which modifications have occurred. These are usually concerned with the effects of genetic background on mutagen specificity (Glover, 1956; Chang, Lennox and Tuveson, 1968; Witkin and Theil, 1960) but examples have also been reported in which a class of mutant has responded specifically to an ancillary treatment, e.g. plating medium (Bridges, Dennis and Munson, 1967; Zetterberg, 1962).

Very little is so far known of the mechanisms which underly such effects. The kinetic differences between the two mutation induction curves in K3/17 do not appear to be the result of trivial causes, and we are left with the idea, stressed by Auerbach and Ramsay (1968), that the curves include components which represent treatment effects on cellular processes which are important in mutation. The altered response in the *cot* strain demonstrates that these cellular events may be influenced by the genetic background.

Since low temperature and buffer act only during irradiation it is tempting to suggest that they act by permitting the induction of more of one type of reversion by u.v. However, the possibility should not be discounted that they may permit the differential damage of some other cellular components which impinge on the mutational pathway at a later stage.

We can be more precise for photorepair. Here the specificity arises from events which occur after u.v. treatment is complete. The obvious explanation for this specificity is that the photoproducts giving inositol revertants are mainly non-photoreactivable. Although this may be true, there is no reason to believe that photoreactivable and non-photoreactivable damages are different photochemically. They are both equivalent in the types of mutant they generate in *Neurospora* (Kilbey and de Serres, 1967), and bacteriophage (Drake, 1966). However, rare instances of mutants produced by intrinsically non-photoreactivable damage would possibly have been missed in these experiments. The *inos* 37401 allele was, surprisingly, the first used to demonstrate that premutational damage is photoreparable in *Neurospora* (Brown, 1951). Although killing was high in these experiments and selective killing cannot be excluded, photoreactivation of mutation does seem to have occurred. Since the strain used was microconidiating and genetically quite different from those used in the present studies, the possibility remains that the photoreactivability of this allele may have been influenced by the genetic background. Photoreactivation may have been prevented in the *cot* and K3/17 strains by inaccessibility of the damage to the repair enzyme rather than the intrinsic non-reparability of the lesions concerned.

These results and those of other workers deal principally with reversion. It may legitimately be asked to what extent they will help us towards the control of mutagen specificity in practice, where we are concerned mainly with forward mutation systems. Mutagens which are specific in reversion tests by virtue of their ability to react with specific damage in DNA often fail to show specificity in forward mutation tests. This is not unexpected since the chemical similarity between genes is such that it is unlikely that one would be singled out by a mutagen. In the same way it could be argued that conditions which modify mutational specificities in reversion experiments are also specific because they recognize specific types of DNA damage. If this is true it may also be unlikely that they would act specifically in forward mutation experiments. In spite of this, however, modification of forward mutational specificities does occur. Zetterberg (1961, 1962) for instance has shown that the plating medium and the mutagen interact to determine the spectrum of amino-acid-requiring mutants in *Ophiostoma*. Amino-acid-requiring mutants are obtained on complete medium after treatment with both nitrosomethylurethane and u.v. However, after u.v., no histidine-less mutants were found although nitrosomethylurethane induced them. If, instead of complete medium, minimal medium plus histidine was used to plate treated cells, histidine-less mutants were obtained after both treatments. It would appear that a constituent of complete medium suppressed the growth of u.v.-induced histidine-less mutants while not inhibiting histidine-less mutants induced by nitrosomethylurethane.

There seem to be two main priorities for mutation research at present: (*a*) the elucidation of the mechanisms by which specificity is modified in reversion experiments, and (*b*) the biochemical investigation of instances, such as the one described by Zetterberg (1961, 1962), in which specificity has been observed in a forward mutational system.

SUMMARY

The u.v.-induced revertibility of two mutants of *Neurospora*, *ad-3A* 38701 and *inos* 37401, is described. The mutational specificity of u.v., expressed as the ratio between inositol and adenine reversions, can be modified by a variety of experimental conditions. U.v. is more specific for inositol reversions at high than at low doses in the K3/17 strain although this effect is dependent on the residual genetic background. Temperature during irradiation, the suspending fluid for the cells and photorepair all act to enhance the i/a ratio. The relationship of these results to the general problem of mutagen specificity is briefly discussed.

Acknowledgements

I am indebted to Dr M. Allison for generously permitting me to quote his unpublished results, and to my colleagues in the MRC Mutagenesis Research Unit for many stimulating discussions.

REFERENCES

Auerbach, C., and Ramsay, D. P. (1967). *Mutation Res.*, **4,** 508–510.
Auerbach, C., and Ramsay, D. P. (1968). *Molec. gen. Genet.*, **103,** 72–104.
Bridges, B., Dennis, R., and Munson, R. J. (1967). *Genetics, Princeton*, **57,** 897–908.
Brown, J. S. (1951). *J. Bact.*, **62,** 163–167.
Champe, S. P., and Benzer, S. (1962). *Proc. natn. Acad. Sci. U.S.A.*, **48,** 532–546.
Chang, L. T., Lennox, J. E., and Tuveson, R. W. (1968). *Mutation Res.*, **5,** 217–224.
Corran, J. (1968). *Molec. gen. Genet.*, **103,** 42–57.
Demerec, M. (1953). *Symp. Soc. exp. Biol.*, **7,** 43–54.
De Serres, F. J. (1964). *J. cell. comp. Physiol.*, **64,** Suppl. 1, 33–42.
Drake, J. W. (1966). *J. Bact.*, **92,** 144–147.
Glover, S. W. (1956). In *Genetic Studies with Bacteria*, pp. 121–136. Washington, D.C.: Carnegie Institution of Washington, Publication No. 612.
Kilbey, B. J. (1963). *Z. VererbLehre*, **94,** 385–391.
Kilbey, B. J. (1967). *Molec. gen. Genet.*, **100,** 159–165.
Kilbey, B. J., and de Serres, F. J. (1967). *Mutation Res.*, **4,** 21–29.
Kølmark, H. G. (1953). *Hereditas*, **39,** 270–276.
Kølmark, H. G., and Kilbey, B. J. (1962). *Z. VererbLehre*, **93,** 356–365.
Malling, H., Miltenburger, H., Westergaard, M., and Zimmer, K. G. (1959). *Int. J. Radiat. Biol.*, **1,** 328–343.
Terry, C. E., Kilbey, B. J., and Branch-Howe, H. (1967). *Radiat. Res.*, **30,** 739–747.
Westergaard, M. (1957). *Experientia*, **13,** 224–243.
Witkin, E., and Theil, E. C. (1960). *Proc. natn. Acad. Sci. U.S.A.*, **46,** 226–231.
Zetterberg, G. (1961). *Hereditas*, **47,** 295–303.
Zetterberg, G. (1962). *Hereditas*, **48,** 371–389.

DISCUSSION

Dawson: In the diauxotrophic strain you score reversions of the inositol allele plated on minimal medium + adenine and reversions of the adenine allele plated on minimal medium + inositol. To what extent does this difference in plating medium affect the results? Do you get comparable results if you plate the single auxotrophs on minimal medium?

Kilbey: The plating medium didn't play a significant role in the results with diepoxybutane.

Auerbach: Plating medium does have an effect in other experiments, as Dr. Clarke has shown (this volume, pp. 17–28).

Bridges: How can you be sure that the difference in medium won't play a role?

Kilbey: This should be tested in each case.

Dawson: A control experiment would be to take the double auxotroph, select a reversion to adenine independence, and then test the inositol reversions, and from the double auxotroph take reversions to inositol independence and test the adenine reversions. Both reversions would then be scored on minimal medium.

Kilbey: Yes, that is right.

Bridges: Is there any reason why you can't use a low level of adenine and inositol so that you can score survival and mutation in the same medium, as we do with almost all bacterial experiments nowadays?

Kilbey: Kølmark and I tried this once. But even so one can never get identical conditions.

Kimball: Did you imply that other inositol mutants, apart from the 37401 you used, were like the adenine mutants and not like 37401?

Kilbey: Yes. The 37401 is so far the only allele which responds to photoreactivation in this way. Fifteen other alleles at three different loci behaved similarly to the adenine.

Kimball: Has their behaviour in other systems, apart from the photoreactivation system, been tested in other respects in which the two differ?

Kilbey: They have not been tested for their response to temperature. Another pair of alleles tested showed no buffer effect, but the data are still very meagre. So far this inositol allele (37401) is the only one which shows the buffer effect. As I mentioned, this was one of the first alleles in *Neurospora* used to demonstrate photoreactivation of reversion (Brown, 1951). These tests were made in a microconidiating strain completely different from ours, which may indicate that the genetic background is important in determining photoreactivability.

Clarke: To what extent is your photoreactivation enzymic photorepair?

Kilbey: As far as we can tell the conditions used give only enzymic photoreactivation. Photoprotection does not seem to occur to a detectable degree.

Clarke: But there might be a photo-physical reversal which wouldn't be mimicked by photoprotection.

Kilbey: This is possible, but the action spectrum for photoreactivation in *Neurospora* shows that little photoreversal was found for the wavelengths at which this might be expected to occur.

Rörsch: Since the reversion to prototrophy of the adenine allele is photoreactivable and that of the inositol allele is not, can it be concluded that the mutated site in the adenine locus contains two adjacent pyrimidines?

Kilbey: I don't know.

Auerbach: It would still be surprising if all the other inositol and adenine alleles did have two adjacent pyrimidines.

Chantrenne: When you use several mutants corresponding to different point mutations in the adenine 3*A* locus do you get the same i/a ratio?

Auerbach: No, but we do not get what we would expect on a simple molecular model.

Chantrenne: You are using two mutants of the same gene, but you do not know whether the two mutations are of the same chemical nature.

Magni: Your conclusions are based on the quantitative evaluation of dose response curves, Dr. Kilbey. What is the variability of these results?

Auerbach: Wouldn't that point be answered by the fact that when one plots all our u.v. data for K3/17 against dose, there is a clear positive regression of the i/a ratio on dose? I shall come to this in my paper (this volume, pp. 66–79).

Magni: I was referring particularly to the temperature experiments. If you repeat the experiments several times do you always get the same result, that is a lower ratio at zero time?

Kilbey: The lowest point is rather variable, but it is always less than 2. At the high dose it is always greater than 2.

Kimball: In all these experiments mutation is of necessity plotted against survival, yet survival doesn't always change in the same way as mutation with various treatments. When you plot the ratio you use the fairly simple, but reasonable, point of view that mutation is independent of survival, in the sense that the surviving sample is a random sample in respect of the mutation process. If this is not so for some reason then changes in survival curve by themselves can give a change in apparent mutation yield.

Kilbey: I think we have to make this assumption. In the present case a constant dose-modifying factor relates the survival curves with and without repair and we therefore conclude that the differential effect of photoreactivation is on inositol reversions.

Dawson: But to what extent might that affect the way one talks about the results? For example, if you say that the reversions of the inositol allele do not respond to photoreactivation are you really saying that they respond to photoreactivation to the same extent as does survival?

Kilbey: I am saying that the adenine reversions are photoreactivable to the same extent as survival; that is why, with and without repair, they fall on the same curve when we plot them in this way. For inositol reversions there is less photoreactivation than for survival.

Auerbach: Mutations are not scored on the same medium as survival, so even in this system where both mutations are scored in the same nucleus, survival may seemingly "create" specificity. I have one example where I think specificity is at least partially dependent on this unavoidable source of error.

Clarke: There exist systems in bacteria where one can use u.v. at non-lethal doses. Shankel and Wyss (1961), for instance, have done this. Mutational systems also exist in which one does in fact score mutations on exactly the same plates as survivals. These are the adenine forward mutation systems in yeast.

Kilbey: I don't know whether one can be entirely sure that there is no lethality in these experiments.

REFERENCES

Brown, J. S. (1951). *J. Bact.*, **62**, 163–167.
Shankel, D. M., and Wyss, O. (1961). *Radiat. Res.*, **14**, 605–617.

ANALYSIS OF A CASE OF MUTAGEN SPECIFICITY IN *NEUROSPORA CRASSA*

CHARLOTTE AUERBACH

MRC Mutagenesis Research Unit, Institute of Animal Genetics, University of Edinburgh

THE system with which I shall be concerned and the techniques used in working with it have already been described by Dr Kilbey (see p. 51). It is the diauxotrophic strain *ad-3A* 38701, *inos* 37401 in which Kølmark discovered mutagen specificity at the same time as Demerec discovered it in bacteria (1953). In a review paper of 1957, Westergaard lists specificities of mutagens in this system, measuring the degree of specificity by the ratio of adenine to inositol reversions, a/i. The lowest ratio in his list, and the only one below 1, is 0·5 for u.v.; this is of course the reciprocal of the i/a ratio of 2 quoted by Kilbey. The most specific agent in Westergaard's list is the difunctional alkylating agent diepoxybutane (DEB); its a/i ratio is given as 445. Dr Kilbey has already shown that these ratios vary and can serve only as a rough guide in comparisons of specificity; how very rough and, indeed, misleading they can be will appear from my report. Nevertheless, I want at this stage to add some equally rough estimates of specificity for three mutagens that I shall have to mention. One is nitrous acid that produces more adenine than inositol reversions, the average a/i ratio being about 10. The others are nitrosomethylurethane and nitrosoethylurethane, both of which produce preferentially inositol reversions, with average a/i ratios below 0·5.

The first instances of mutagen specificity were discovered just before the beginning of the Crick-Watson era; they could not therefore be linked with well-founded hypotheses about the chemical reactions between genes and mutagens. Nevertheless, all speculations about the origin of mutagen specificity already centred around what one may call the primary step in mutagenesis and, on the whole, this has remained so. To me, this has always seemed an unbiological point of view. Apart from experiments on virus treated *in vitro*, we always treat whole cells. It is most unlikely that the mutagens which we use should act exclusively on chromosomal DNA. Indeed, there is now ample evidence to show that ionizing radiation, u.v., alkylating agents and other mutagenic chemicals affect cellular processes that form part of the mutagenic pathway. Long before this knowledge was available, it seemed probable to me that mutagen specificity might arise from the action of mutagens on secondary processes in mutagenesis. This idea has been my guideline in the analysis of Kølmark's historic case.

Somc of the early experiments have been done in cooperation with Dr Kølmark, most of the later ones in cooperation with Mr D. Ramsay. It is important to realize that we have studied reverse mutations, very few of which appear to be due to extracistronic suppressors. A distinction is generally made between mutagen specificity for forward and reverse mutations. For the former, it is said, mutagen specificity is not to be expected because a gene can be damaged in many different ways. Obviously, this is true. Nonetheless, there have been reports of quite striking mutagen specificity for forward mutations; Dr Kilbey has mentioned one of them. In all these cases specificity is probably created by the setting-up of internal "sieves" that, in treated cells, inhibit the expression of certain potential mutations and favour that of others. In reverse mutations, one particular previously established damage to DNA has to be reversed. This, it is said, provides a basis for mutagen specificity. Again, there is truth in this contention but, for it to be verifiable, one has to work with a system in which true reversals can be separated from suppressors, including intracistronic and intracodon ones. Moreover, there is no reason why those cellular processes that act as sieves for forward mutations should not do the same for reverse ones and thus create specificity over and above that due to specific reactions between DNA and mutagen. Indeed, internal sieves should be particularly effective in screening out potential revertants because the screening procedure for mutations from auxotrophy to prototrophy sets a time limit within which a potentially mutant cell has to achieve biological competence. The results to be reported here show that, in Kølmark's reversion system, specificities arising at cellular levels are an important, perhaps the most important, source of mutagen specificity.

Much of our work has been concerned with the most specific mutagen in Westergaard's (1957) list, DEB, which not only produces extremely high frequencies of adenine reversions but also very low ones of inositol reversions. From what we now know about the reactions of alkylating agents with DNA, this might be explained by the assumption that reversal of the adenine allele involves reaction with guanine, while reversal of the inositol allele does not. There are, however, reasons for doubting this simple interpretation. First, DEB does produce inositol reversions, although at low frequencies. Second, Dr Allison found (1969*a*) that the inositol allele responds very well to another alkylating agent, ethyl methanesulphonate; a second inositol allele showed the same reactions: refractoriness to DEB coupled with responsiveness to ethyl methanesulphonate. In contrast, of four *ad-3A* alleles tested three responded well to both alkylating agents, although to different degrees, while the fourth responded to neither. This last may have been one of the cases in which a particular type of damage in DNA cannot be reversed by alkylation.

In any event, these findings were not known to me when I embarked on a search for cellular causes of the specificity of DEB. Trivial causes, such as differential killing of revertants or differential sensitivity to crowding on the plates, had already been ruled out by Dr Kølmark or were ruled out later by Mr Ramsay and myself. The first possibility that came to my mind was that DEB treatment might sensitize potential mutations to the differences in plating medium which are an unavoidable feature of reversion experiments with diauxotrophic strains. Dr Clarke's experiments (1962) have shown that such differences can indeed give rise to mutagen-specific responses. In our system, however, this source of specificity could be ruled out by experiments of the kind described by Clarke.

I then decided to carry out an experiment which should show whether DEB affects any one of the steps in the two mutational pathways or whether it has no such effect at all. I argued that, if treatment of a cell with DEB by any means whatever facilitates the realization of potential adenine reversions and/or inhibits the realization of potential inositol reversions, then this cellular effect might persist into a subsequent treatment with a different

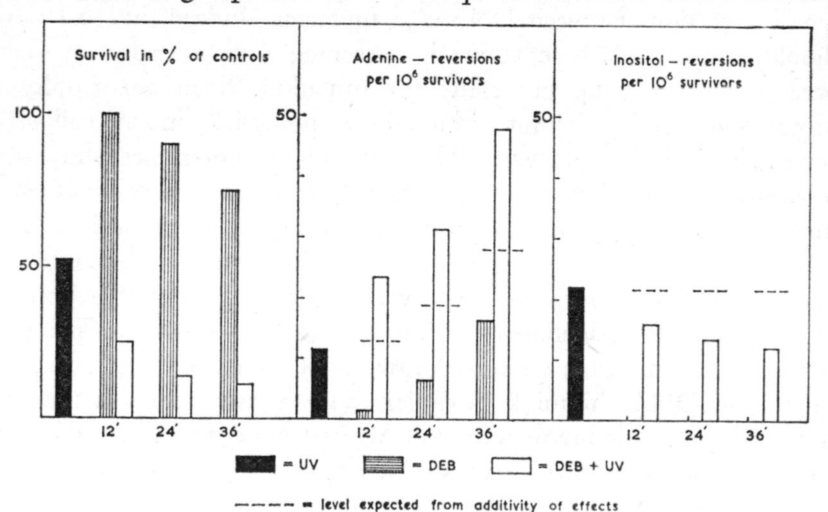

FIG. 1. Combination treatment: DEB followed by u.v.

mutagen and modify *its* specificity in the direction of that found for DEB. U.v. was, at that time, the only known mutagen that produced more inositol than adenine reversions and we compared its effect in cells that had or had not been pretreated with DEB. This experiment has been repeated many times, and the result was nearly always the same. The few exceptions were due to unusually high doses of DEB.

An example is shown in Fig. 1. In every series pretreatment with DEB—even with a dose that was hardly mutagenic—increased the frequency of adenine reversions beyond additivity and decreased the frequency of inositol reversions below additivity. Since survival in this particular

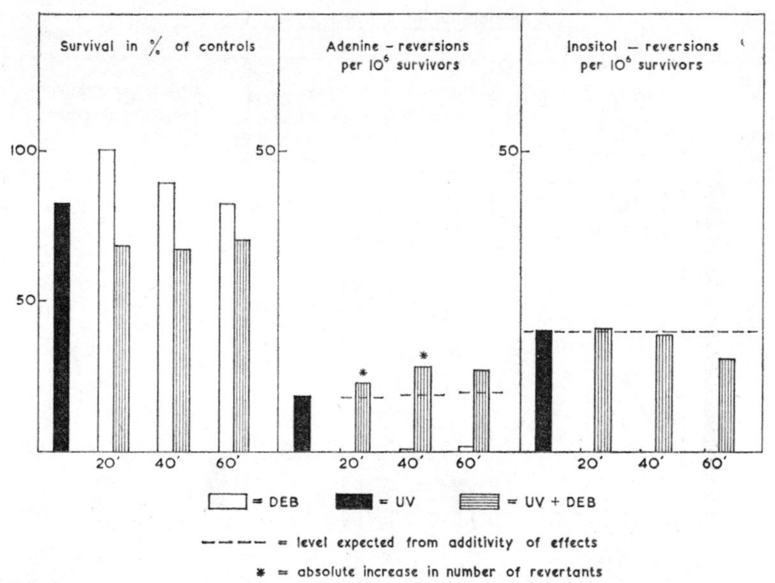

FIG. 2. Combination treatment: u.v. followed by DEB.

experiment was low, there was no absolute increase in the colony counts of adenine reversions, but this has been obtained repeatedly in experiments with higher survival, as can be seen in Fig. 2, which presents results from an experiment in which u.v. was followed by DEB. This type of experiment gives variable results, depending on the relative strengths of the two treatments. Enhancement of adenine reversion frequency requires that the dose of DEB be mutagenically less effective than the preceding u.v. dose. In this condition, enhancement is regularly observed and may be absolute. It is not destroyed by an hour's intermediate incubation in water.

Other chemicals with preferential action on the adenine allele give the same type of interaction with u.v. Fig. 3 shows an experiment in which nitrous acid was given either as pre- or post-treatment: in both cases, the frequency of adenine reversions was enhanced and that of inositol reversions was reduced in relation to what would have been expected from additivity of the treatment effects. Malling and co-workers (1959) have shown that even a barely mutagenic substance like formaldehyde can affect u.v.-induced reversions in exactly the same way. I find it difficult to avoid the conclusion that all these treatments, whether or not mutagenic in their own right, and whatever the mechanism of their action on DNA, create conditions in the cell that favour the appearance of adenine reversions and inhibit that of inositol reversions.

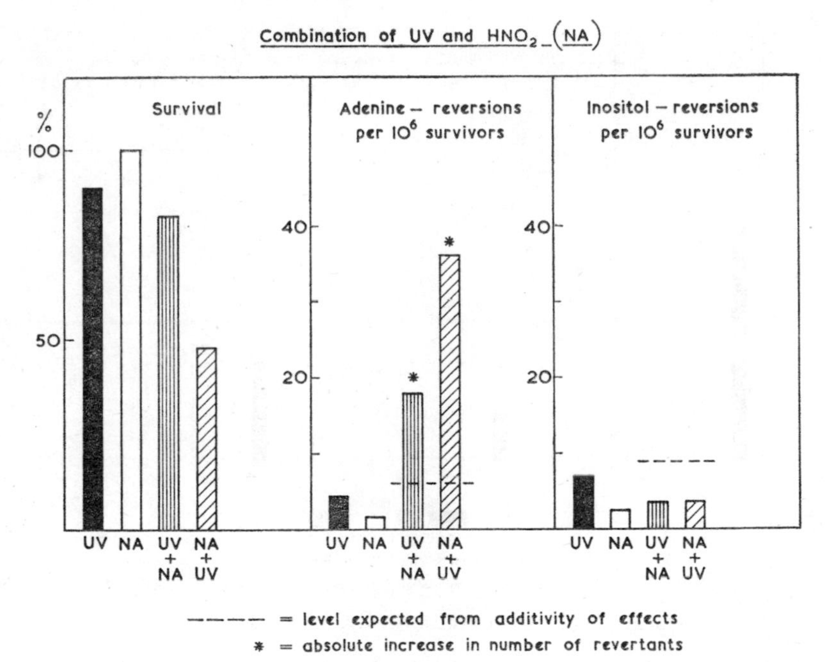

FIG. 3. Combination treatment: nitrous acid (NA) followed or preceded by u.v.

The rationale behind the interaction experiment had been the idea that DEB yields many adenine reversions and few inositol reversions because, in addition to producing an unknown proportion of mutagenic changes in the DNA of both alleles, it affects cellular processes that interfere in different ways with the mutagenic pathways. When the results were in accordance with expectation, we put this idea to a more direct test. The interaction experiments had already shown that very low doses of DEB can act as "boosters" for the production of adenine reversions by u.v. We now tested whether they could also act as boosters for adenine reversions produced by DEB itself. Fig. 4 shows that this is indeed the case. A dose of normal mutagenic strength produces many more adenine reversions in cells that have been pretreated with a very weak, barely mutagenic dose than in cells that have not been so pretreated.

Fig. 5 shows that the same result can be obtained when a high dose of DEB is followed by a very low one. Here the high dose has not only produced a considerable number of mutations; it has also sensitized the

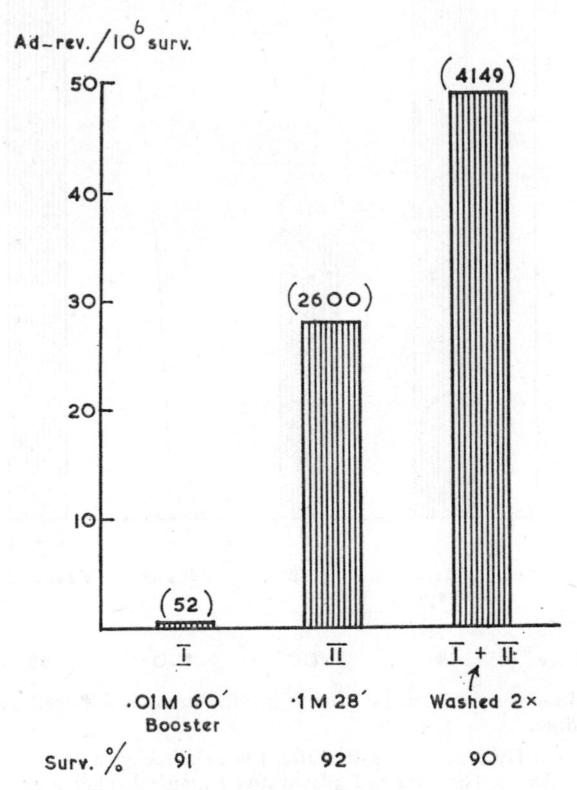

FIG. 4. Booster experiment: a weak booster dose of DEB followed by a dose of normal strength. The figures in parentheses represent the number of colonies counted.

cells to the subsequent effect of a very low dose, which by itself produces hardly any mutations in fresh conidia. This experiment provides an explanation for the so-called after-effect of DEB, shown on the left of Fig. 5. When DEB-treated, twice-washed cells are kept standing in water, adenine reversions continue to arise for many hours. Kølmark and Kilbey (1962) found that this is due to traces of DEB which can be removed from the cells only by many hours of washing in stirred liquid. Yet, when these

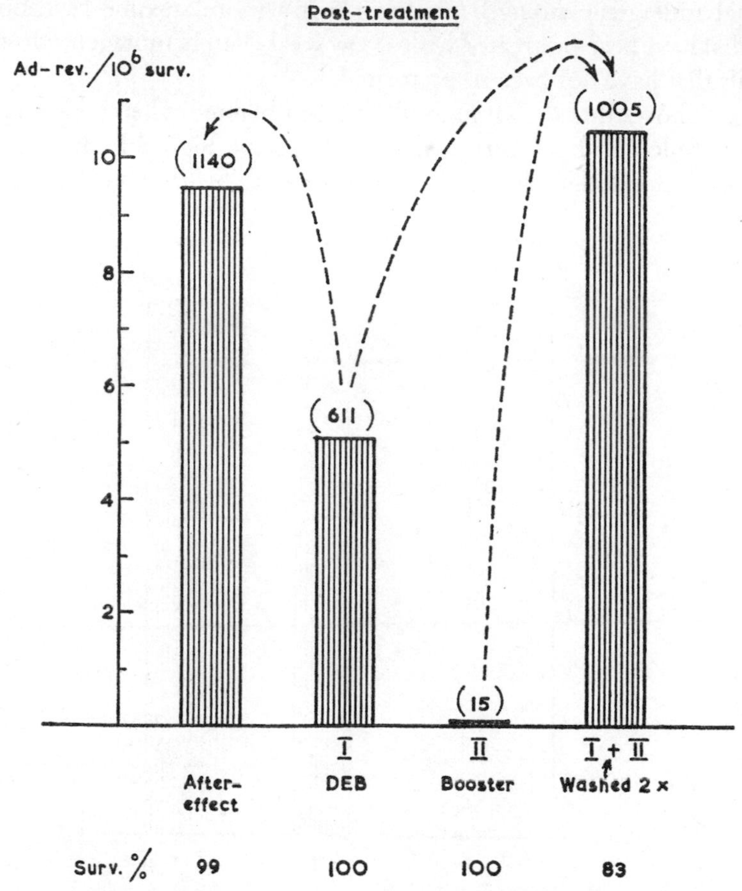

FIG. 5. Booster experiment: a dose of normal strength followed by a weak booster dose.

I: 0·1 M-DEB, 22′, 25°, plated after 1 hour's shaking at 25°.
After-effect: The same as I, plated after 1 hour's shaking at 25°, followed by spinning down and 4 hours' standing at 25°.
II: 0·01 M-DEB, 60′, 25°, plated at once.
The figures in parentheses represent the number of colonies counted.

traces had finally been removed, the supernatant failed to produce mutations in fresh conidia. Although the conditions in this experiment were different from those giving an after-effect, the most plausible explanation is that fresh conidia do not respond to the very small amounts of DEB to which previously treated cells have become sensitized.

Clearly, this dual action of DEB—production of potential adenine reversions and sensitization of the cell to their realization—must generate a steep dose-effect curve. This is indeed true. Kølmark and Kilbey (1968) fitted their results to the usual equation $m = a + bd^x$, where m = mutation frequency, d = dose, a and b = constants, and they found a good fit to $x = 2{\cdot}5$. I obtained the same dose-exponent for the pooled results from my experiments (Fig. 6). This exponent is considerably higher than would be

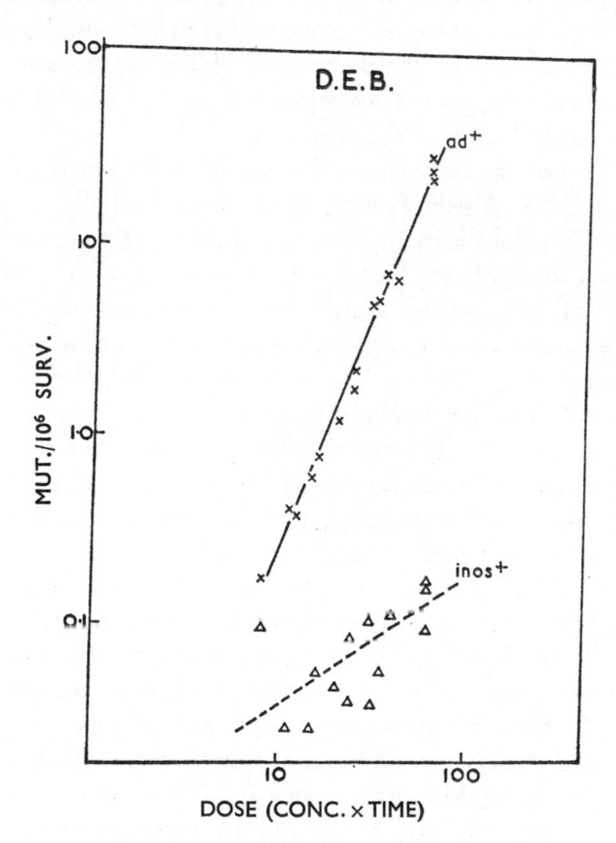

FIG. 6. Dose-effect curve for adenine reversions (ad^+) and inositol reversions ($inos^+$) after treatment with DEB. The ordinate shows mutation frequencies per 10^6 survivors.

(Reprinted from Auerbach and Ramsay, 1968*a*, by permission of Springer.)

expected if DEB produced reversions by a two-hit mechanism; anyway, there is no reason to assume that it does so. To me, the most likely interpretation is that the observed dose-effect curve is a composite one in which the presumably linear relationship between amount of DEB and alkylations in DNA is overlaid by steeper dose responses of cellular steps in the mutational pathway. There is some independent evidence for this interpretation. Kølmark and Kilbey (1968) found that the dose-effect curve for adenine reversions becomes linear when DEB is administered at low concentrations over long periods. This is reminiscent of the dose-rate effect of X-rays on mice (Russell, Russell and Kelly, 1958) and suggests that DEB given at very low dose rates has only a negligible action on cellular processes. It is interesting that, years ago, Novick (1955) made a similar observation on bacterial mutations in the chemostat. After exposure to acute u.v. irradiation, the dose exponent lay between 2 and 3; after chronic exposure to low-incident radiation, it dropped to 1.

Dose rate is not the only factor that modifies the kinetics of reversion induction by DEB. Allison (1969*b*) tested three *ad-A* alleles in combination with two inositol alleles; in every case, the dose-effect curve for adenine reversions was steeper in the background of *inos* 37401 (the allele we used) than in that of *inos* 37102. The kinetics of survival paralleled that of reversion, being steeper in strains with 37401 than in strains with 37102. It seems that the two inositol alleles—or alleles of a linked gene—influence the response to DEB of some cellular process that affects killing and mutation in the same sense. On the simplest interpretation, interference with this process also causes the other features of DEB-induced adenine reversions, i.e. the after-effect, the booster effect, and the enhancement of u.v.-induced reversion frequency by pre- or post-treatment.

The nature of this process is a matter for speculation. One suggestive fact emerged (Auerbach and Ramsay, 1968*a*) when the findings from our experiments were plotted on a semi-log scale (Fig. 7). It then appeared that the adenine reversion curve has a strong exponential component of the form $m = a\,e^{cd}$, where m = mutation frequency, d = dose, e = base of natural logarithms, and a and c are constants. In fact, when I plotted Kølmark's original results on semi-log paper, they fell almost exactly on a straight line. This suggests that one of the cellular processes in the mutagenic pathway reacts exponentially to treatment with DEB. Inactivation of a repair enzyme would fit in with a number of observations, including the parallel effect of genetic background on survival and mutation, the linear dose-effect curve at very low dose rates, and the enhancement of u.v.-induced adenine reversions, even when DEB is given as post-treatment. Unpublished results of Dr Kilbey's suggest that in yeast DEB inactivates repair of colony-forming ability. None of these findings provide conclusive evidence on this point, and the possibility remains that the peculiarities of DEB-induced adenine reversions are based on interactions with other steps than repair in the mutational pathway.

By whatever mechanism DEB sensitizes the cell to adenine reversions, it cannot have the same effect on inositol reversions, for these behave in every way as the opposite of adenine reversions. Their overall frequency is very low; their dose-response curve is very flat, with a dose

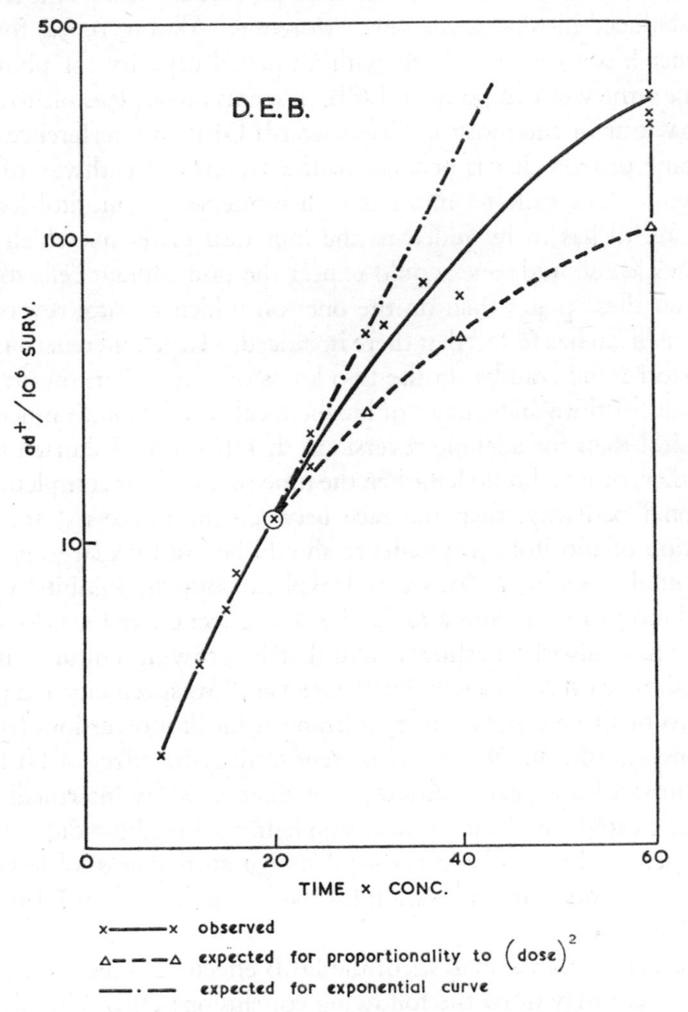

FIG. 7. Dose-effect curve for adenine reversions on a semi-logarithmic scale.

exponent of less than 1; they show no after-effect and no response to booster; and when u.v. irradiation is followed or preceded by DEB, the frequency of inositol reversions is, at best, unchanged but is usually reduced. If one ascribes the peculiarities of the adenine response to inactivation of a repair enzyme, one would have to assume that the inositol allele does not respond to this enzyme. Kilbey's (1967) finding that the same inositol allele is refractory to photorepair lends some plausibility to this assumption. It cannot, however, account for dose exponents of less than unity nor for the fact that pre- or post-treatment with DEB actually *reduces* the frequency of u.v.-induced inositol reversions. Moreover, Allison (1969*a*) found that two other inositol alleles, both with a normal capacity for photorepair, show the same weak response to DEB. It seems more plausible to attribute the behaviour of the inositol allele towards DEB to interference of DEB with some process that is peculiar to the mutational pathway of inositol reversions. This calls to mind the phenomenon of inositol-less death. Since inositol has to be added to the mutation plates on which adenine reversions are scored, one would expect the non-mutant cells to survive longer on these plates than on the ones on which inositol reversions are scored. Allison has found that there is, indeed, a large difference in survival of diauxotrophic conidia on the two kinds of plate. This means that the time limit for the achievement of biochemical competence is much shorter for inositol than for adenine reversions. If DEB should shorten this time still further, or if it should lengthen the time required for completion of the mutational pathway, then the race between inositol-less death and the acquisition of inositol independence should be lost by a large proportion of potential revertants. This could explain both the inhibition of u.v.-induced inositol reversions and the flat dose-effect curve for DEB-induced ones. It may also be pertinent here that in growing cultures, in which inositol-less death is not a risk, DEB has a very low specificity and produces only two or three times as many adenine as inositol reversions (Auerbach and Ramsay, 1968*b*). We are at present testing the effect of DEB on the rate of inositol-less death. Attempts to stave it off by intermediate incubation of treated conidia in inositol-supplemented medium did not increase the frequency of inositol reversions; this suggests that inositol-less death by itself is not a sufficient explanation for the low frequency of DEB-induced inositol reversions.

From these separate analyses of the DEB effects on adenine and inositol reversions we may draw the following conclusion: DEB is highly specific in this system because it produces adenine reversions by a self-enhancing action and inositol reversions by a self-inhibitory action. This leads one to expect that the degree of specificity should increase with dose. Fig. 6 shows that this is indeed true. At the lowest doses the ratio between adenine and inositol reversions was less than 10, at the highest doses it was more than 200. The ratio given by Westergaard (1957) was even higher; possibly it

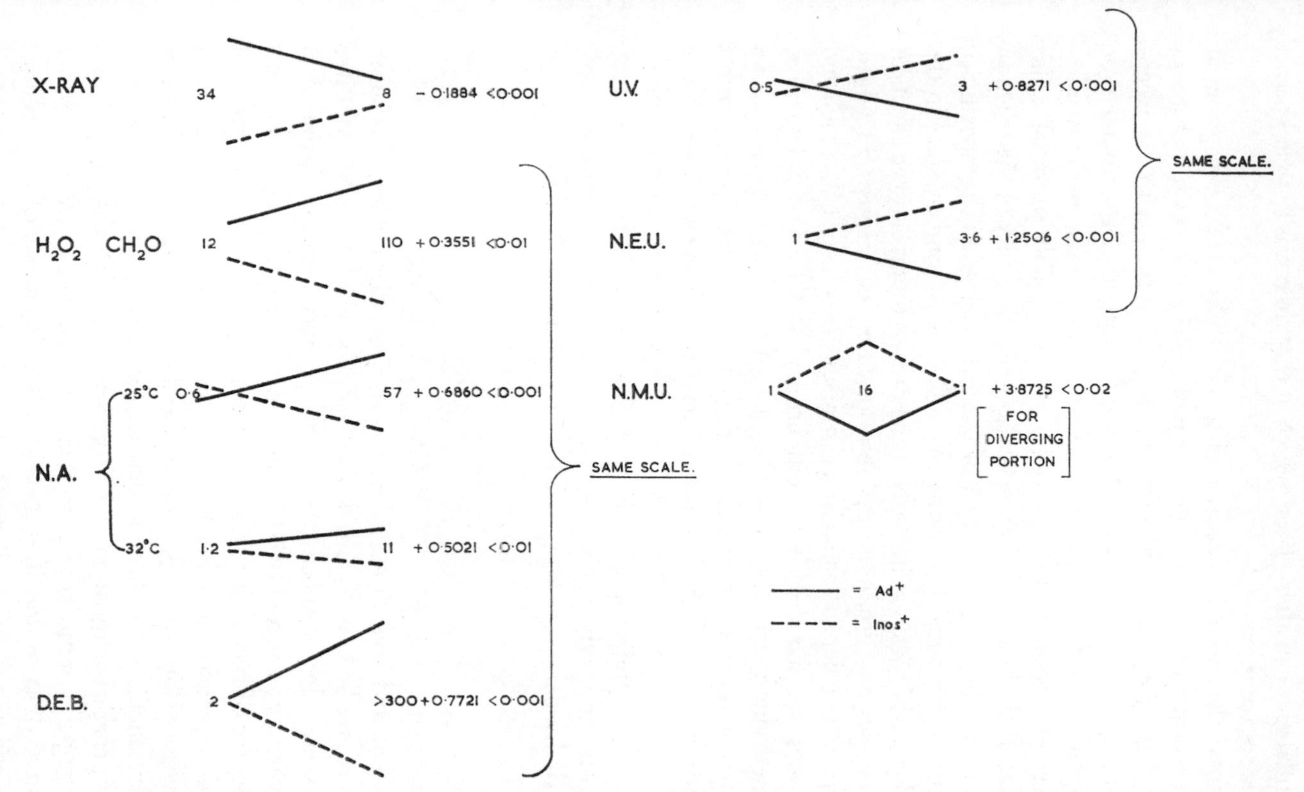

FIG. 8. Summary of data showing the dose dependence of mutagen specificity. a: frequency of *ad*⁻ reversions per 10^6 survivors; i: frequency of *inos*⁻ reversions per 10^6 survivors. The lines do not represent dose-response curves, but changes in specificity with dose. Divergence means that specificity increases with dose; convergence, that it decreases. The numbers to the left and right of each pair of lines are the extreme values of the a/i ratio (left) or i/a ratio (right). b: regression coefficient. Note that regression coefficients can be compared only when they have been derived from data calculated on the same scale.

NA: nitrous acid
NEU: nitrosoethylurethane
NMU: nitrosomethylurethane

(Reprinted from Auerbach and Ramsay, 1968*a*, by permission of Springer.)

had been inflated by an after-effect, which at that time was not yet known as a possible source of error.

Fig. 8 shows that dose dependence of specificity is not an exception, at least not in Kølmark's system. Every one of eight mutagens tested shows dose-dependent specificity. The most striking fact is that, with the exception of X-rays and, perhaps, the radiomimetic mixture of formaldehyde and hydrogen peroxide, specificity is absent or very weak at the smallest doses, where effects on cellular processes are presumably minimal. This suggests that in all these cases there is little or no selectivity in the reaction between the mutagen and the DNA of the two alleles and that specificity becomes imposed gradually as treatment effects on secondary steps in the mutational pathways gain in importance. For two mutagens, the effects of cellular processes on specificity are obvious. One is nitrous acid whose degree of specificity depends on temperature *after* treatment (Auerbach and Ramsay, 1967); this has already been mentioned by Dr Kilbey. The other is nitrosomethylurethane. Here, divergence of the dose-response curves gives way to convergence at high doses; the turning point coincides almost exactly with the dose at which survival starts to fall precipitously. Since a reconstruction experiment showed that high doses of nitrosomethylurethane do not preferentially kill completed inositol revertants, they must discriminate against potential ones, possibly by speeding up inositolless death or by delaying some step in the pathway to inositol independence.

SUMMARY

X-rays, u.v. and five different chemicals (not all of them alkylating agents) were applied to Kølmark's *ad⁻inos⁻* strain. X-rays and three chemicals were adenine-specific, i.e. they produced more adenine than inositol reversions; u.v. and two chemicals were inositol-specific. In every case, the degree of specificity was dose-dependent; with the exception of X-rays, it always increased with dose and was weak or absent at the lowest doses. Reversion kinetics and, with it, the degree of specificity could be modified by ancillary conditions such as dose rate, post-treatment temperature or genetic background. Diepoxybutane, the most highly adenine-specific mutagen, lost most of its specificity when it was given to a growing culture instead of to conidia in suspension. The frequency of u.v.-induced adenine reversions was increased and that of inositol reversions was reduced when irradiation was preceded or followed by weak treatment with an adenine-specific chemical.

These observations have led us to the conclusion that most of the observed mutagen specificity in this system is due, not to specific chemical reactions between mutagen and DNA, but to treatment effects on cellular processes that play different roles in the mutational pathways of the two alleles. There are indications that inactivation of a repair enzyme and hastening of inositol-less death may both be involved.

REFERENCES

ALLISON, M. (1969*a*). *Mutation Res.*, **7**, 141–154.
ALLISON, M. (1969*b*). *Mutation Res.*, in press.
AUERBACH, C., and RAMSAY, D. (1967). *Mutation Res.*, **4**, 508–510.
AUERBACH, C., and RAMSAY, D. (1968*a*). *Molec. gen. Genet.*, **103**, 72–104.
AUERBACH, C., and RAMSAY, D. (1968*b*). *Jap. J. Genet.*, **43**, 1–8.
CLARKE, C. H. (1962). *Z. VererbLehre*, **93**, 435–440.
DEMEREC, M. (1953). *Symp. Soc. exp. Biol.*, **7**, 43–54.
KILBEY, B. J. (1967). *Molec. gen. Genet.*, **100**, 159–165.
KØLMARK, H. G. (1953). *Hereditas*, **39**, 270–276.
KØLMARK, H. G., and KILBEY, B. J. (1962). *Z. VererbLehre*, **93**, 356–365.
KØLMARK, H. G., and KILBEY, B. J. (1968). *Molec. gen. Genet.*, **101**, 89–98.
MALLING, H., MILTENBURGER, H., WESTERGAARD, M., and ZIMMER, K. G. (1959). *Int. J. Radiat. Biol.*, **1**, 328–343.
NOVICK, A. (1955). *Brookhaven Symp. Biol.*, **8**, 201–215.
RUSSELL, W. L., RUSSELL, L. B., and KELLY, E. M. (1958). *Science*, **128**, 1546–1550.
WESTERGAARD, M. (1957). *Experientia*, **13**, 224–234.

DISCUSSION

Dawson: You showed the dose of diepoxybutane (DEB) as a product of concentration and time, yet if a prior small dose boosts the frequency of mutations induced by the main dose, more mutations may be induced with a low concentration for a long time than with a higher concentration for a short time. Wasn't it therefore unsatisfactory to plot the dose as a product of time and concentration?

Auerbach: In fact the opposite is true. Except for very low concentrations, the product of time and concentration gives an excellent fit to a dose-effect curve with exponent 2·5. Long exposure to a very low concentration gives a linear curve.

Dawson: Do you normally wash the cells after giving the booster dose and before giving the main dose?

Auerbach: Yes.

Dawson: There isn't formally very much difference between giving a certain concentration for a long time and giving two separate doses, except the interval of washing in between. Therefore this period appears to be critical for DEB sensitizing the cells.

Auerbach: I have also given the booster without washing; the result is the same.

Dawson: Then why should this be different from just giving DEB for a longer period?

Auerbach: It isn't. If I give a large dose which produces, say, 600 mutations, and then a very small one which produces 50, I may obtain 900 mutations, but I could also get 900 by giving the high dose for a few minutes longer.

Loveless: The question might be resolved when you know what happens if you give a small dose and then another small dose.

Auerbach: Presumably the same, but I could try.

Apirion: Is the washing so effective that the DEB is completely removed?

Auerbach: Kølmark and Kilbey (1962) found that DEB could be removed completely by shaking the cells for several hours in liquid.

Kilbey: We don't really know whether it is DEB or some derivative that causes the after-effect. The supernatant derived from treated cells after prolonged shaking is not mutagenic in the adenine reversion system.

Kimball: If you give a high concentration for a single period of time, sample, and then give it for a slightly longer period of time and sample again, do you get the same general increase that you would expect if you had given a very small dose first, followed by washing and then the larger dose? In other words do you need the washing in between? I think this would answer the point about what happens with two small doses one after another.

Pollock: Your interpretation tends to emphasize adenine and inositol metabolism as an explanation of the difference. It might depend on what type of mutation occurs within the adenine and inositol loci. Would you expect to find the same effect with other mutations at these loci?

Auerbach: No. This explanation emphasizes inositol metabolism. All three inositol alleles we tested showed the same less-than-linear dose-effect curves with DEB. But adenine-3A reversions showed varying responses to DEB.

Pollock: So you have tested more than one mutant?

Auerbach: Only for the dose-effect curves. Dr Allison found that three inositol reversions hardly responded to DEB at all, but all three responded well to ethyl methanesulphonate. The adenine reversions all responded to DEB, but two of them did not do so as strongly as the third.

Magni: The correlation of dose with concentration × time does not hold for other mutagens in bacteria and yeast. For example with nitrosoguanidine one can have more or less the same curve against time with doses differing by a factor of 10. In *Neurospora* with ICR compounds, as observed by H. Malling (personal communication) at any dose, after treatment for a certain length of time, the number of mutants levels off to a plateau depending on the dose. With such an agent one cannot assume that dose equals concentration × time.

Auerbach: It is simply a fact that time by concentration is a satisfactory measure of dose for DEB. For nitrosomethylurethane and nitrosoethylurethane it is not.

Apirion: It is impossible not to have a dose effect; however it could seem to be absent if the lowest tested dose already saturated the system.

Magni: I am not saying there is no dose effect, but that with nitrosoguanidine the dose-effect curve is different from the time-effect curve. If the two curves are different one cannot assume that dose is equal to concentration × time.

Loveless: We are of course talking about substances which are highly unstable in water.

Brookes: DEB is difunctional and the other compounds are monofunctional. Is there any significance in this fact? For example, do you get any delayed effect after you have removed this agent? These experiments with DEB, particularly where you give a high dose and then a low dose and get enhancement, are difficult to understand.

Auerbach: Kølmark and Kilbey (1968) got the same after-effect with ethylene oxide, which is a monofunctional analogue of DEB.

Brookes: When you gave 0·1 M-DEB, removed it by washing and then compared the adenine-inositol ratio to survivors with incubation time thereafter, did the ratio change with time and did survival change with time?

Kilbey: If one permits the after-effect to continue there is a continued increase in mutations and later a drop in survival occurs. In the case of ethylene oxide the times allowed for the mutagenic after-effect were too short for survival to decrease.

Brookes: Does post-treatment incubation lead to an increased a/i ratio?

Kilbey: Yes; there does not appear to be an after-effect on inositol reversion.

Brookes: After you have given 0·1 M-DEB and have washed the cells, have you tried treatment with buffer, instead of the low booster dose? Time is perhaps the important factor.

Auerbach: No, it is not. During the booster treatment, the cells are shaken, and this destroys the after-effect. For every booster experiment, I have a control in which conidia of the same treated and washed sample are shaken in water during the period of the booster treatment; this gives no increase in the frequency of adenine reversions.

Dawson: Do you recognize any mosaics after using DEB?

Auerbach: No, they cannot be recognized in this system.

Apirion: What is the duration of the DEB effect? How long can one wait between applying DEB and nitrous acid and still get the enhanced effect?

Auerbach: I haven't yet done it with nitrous acid. With DEB it has to be done under conditions which remove the after-effect. It is then found that interaction and after-effect both become weaker with time. I interpret this to mean that the sensitizing effect on the pathway to adenine reversions persists for the same length of time, and that it makes no difference whether the booster is ultraviolet or traces of DEB.

Loveless: Dr Kilbey said that the supernate he obtained after shaking his DEB-treated material under conditions which abolished the after-effect did not itself prove mutagenic. Did you test whether it had a booster effect, which would be a more sensitive test of whether anything was there?

Auerbach: No, but this is a good idea, which should be tested.

Kilbey: Can you add the DEB very quickly after u.v.?

Auerbach: Yes, immediately. However, in a few experiments, I kept the irradiated cells before adding DEB, and for up to one hour this did not affect the result.

Clarke: Can you abolish the DEB after-effect just by keeping the conidia cold, or do you wash them on membranes?

Auerbach: I don't think one abolishes the after-effect by keeping them cold. It is just kept in abeyance.

Kilbey: Membrane washing has the same effect as centrifugation on the after-effect: the chemical on the outside of the cell is removed, but something is left inside the cell which produces the after-effect.

Clarke: Does thiosulphate abolish the after-effect?

Kilbey: No.

Apirion: In order to find out whether the after-effect is due to transcription and/or translation of the mutated gene it might be useful to perform similar experiments with *Bacillus subtilis*, for instance. Since *B. subtilis* can be treated, the DNA can be extracted at any point thereafter, and its biological activity tested. Thus the point when mutated DNA is transferable can be assessed.

Auerbach: This is the technique used by Professor Kaplan. We are hoping to use extracellularly treated phage and give ancillary treatment to the host in order to distinguish between direct effects on DNA and effects on later steps in mutagenesis.

Kaplan: It is certainly important to distinguish between the specificity at the DNA level and the higher levels. In our system we could show with extracellular irradiation (which is at the DNA level) strong differences between different types of mutations in the same phage: the *c*-mutations and the *b*-mutations go up with the X-ray dose according to one-hit curves, but the *e*-mutations go up with the third power. But didn't you mean that the DNA level is not very important?

Auerbach: Certainly there must be specificity at the DNA level; but in our system, which is the one always quoted as an example of specificity, it turns out that most, if not all the specificity, occurs at levels other than interaction with DNA.

Hütter: You said you got this specific effect with three inositol alleles. The leakage out of pools might have a significant effect on inositol-less death, which would increase. The observed specificity may have nothing to do with the alleles or with the DNA but with increased leakage followed by cell death.

Auerbach: We did not check leakage. If diepoxybutane increased the leakiness of inositol from the pool, this could explain why it doesn't produce inositol reversions: they would die from inositol-less death before the mutation is expressed.

This is a question of definition. We never see what actually happens in DNA. We only see recovered mutations, and the specificities which we get may or may not arise in DNA. Even if they arise by the death of potential mutants they still appear as specificities, although their source is physiological. Selection against completed mutants can be ruled out by reconstruction experiments.

Hütter: But you could escape that difficulty by plating the strains on media where you do not get inositol-less death.

Auerbach: I have given them intermediate incubation in inositol before plating for reversions. You can't do this for more than one or two hours because then the auxotrophs overgrow the plates. Under these conditions there was no better response to DEB.

Brookes: Before we discard primary alkylation of the DNA as capable of offering an explanation of the reversion phenomenon, let us assume that adenine reverts by an alkylation of guanine, whereas the inositol reverts by alkylation of adenine and subsequent depurination. Two very different lesions result which may be acted on differently by the repair system. The repair system may enhance one reversion while inhibiting the other. The treatments which seem to enhance adenine reversion are ones which might be expected to inhibit repair.

Auerbach: The very strong exponential component has made me consider the possibility that repair is involved. The inositol locus responds very well to ethyl methanesulphonate, but shouldn't this compound act in the same way as diepoxybutane? After treatment with nitroso compounds, inositol reversions were more frequent than adenine reversions. For all the agents that I showed here, except for X-rays and the mixture of hydrogen peroxide and formaldehyde, specificity is absent or very weak at low doses. For all these it increases with dose, so that the a/i or i/a ratio shows a clear regression on dose.

Pollock: How did you get your three different inositol-less mutants in the first place? They might all be of a certain type.

Auerbach: They may be, but all of them give the same low response to DEB. They differ from each other not only in their responses to photorepair (Kilbey, this volume, pp. 50–62) but also in their responses to u.v. and ethyl methanesulphonate (Allison, 1969).

Maaløe: In experiments of this kind it would be nice to be able to test the cells at various times for the presence of the genetic material we are interested in, or its product. Thus with a sufficiently sensitive transformation system the DNA extracted from the cells at a given time could be assayed for the presence of biologically active DNA of the revertant type.

Alternatively, it would in some cases be possible to assay individual cells for the early appearance of the enzyme corresponding to the revertant genotype. If the cells are held in buffer waiting for an after-effect, one could give them 5 minutes in a rich medium, then kill them and see whether during that short time they had produced the enzymes whose synthesis would require a completely reverted genome.

Auerbach: Yes. I think for that type of work one has to go over to bacteria.

Apirion: It is not as clear-cut as it seems. If there is DNA with a lesion, will it penetrate the cell? If it does, then this DNA will be subjected to host cellular processes and therefore the mutation observed would actually become fixed in the recipient cell.

Kimball: I think the point that Dr Apirion is trying to make is that some of the intermediate processes leading to mutation are simply not detectable in any procedure we now know.

Maaløe: My proposal, although technically very demanding, was more modest. I wanted only to define the time, after the mutagenic treatment, at which "fixation" had occurred.

Auerbach: Could one not test when, say, in a super-suppressor, the transfer RNA has appeared?

Maaløe: That is a very difficult assay. I was thinking of products as easy to detect as alkaline phosphatase or β-galactosidase, where killed cells can be made permeable to a substrate which, if acted upon by the enzyme in question, will cause the cell to fluoresce.

Magni: Dr Kilbey and Dr Auerbach have been talking about one kind of mutagen specificity, but we could also take into account other types such as the real molecular specificity at the DNA level. Of course Dr Auerbach's specificity is not molecular specificity in terms of specific chemical changes induced by the mutagen in the polynucleotide sequence. We should at least try to call these two specificities, let us say "DNA specificity" and "cellular specificity", although this does not satisfy me completely.

Auerbach: What is the difference operationally?

Evans: You are discussing specificity at the phenotypic level, that is a heritable change observed at the phenotypic level and which affects the expression of a given allele, Professor Auerbach. Professor Magni's specificity is a molecular specificity not just at the DNA level, but at the nucleotide level. Mutational specificity at the phenotypic (allelic) and nucleotide levels should be clearly distinguished.

Auerbach: Allele specificity does not necessarily arise at the level of DNA. In fission yeast, for instance, different alleles of the same adenine gene respond differently to methionine suppression. In an ad^{-} met^{-} strain, this creates allele specificity (Clarke, 1962). Site specificity in phage is really DNA specificity.

Magni: Operationally the best way of defining the two types is chemical specificity and phenotypic specificity. You are looking for phenotypic specificity of unknown chemical origin, Professor Auerbach, and what we call specificity is chemical specificity.

Auerbach: It is very difficult to draw a borderline operationally. There may be one type of DNA specificity which specifically responds or does not respond to a repair enzyme, so that what one observes has already been subject to a cellular process. Operationally one can draw the distinction only in, e.g., extracellularly-treated phage, or in a system like that of Yanofsky (1968), where the mutant protein can be analysed.

Magni: Or one could use other systems where one can detect the molecular type of mutation.

Auerbach: Yes, but only whole classes like base changes or frame shifts, not whether the base change was from A to G or from G to A.

Loveless: This whole problem of the multiple use of the expression "specificity" was dealt with admirably by Professor M. Westergaard in 1959 at the first Gatersleben symposium on chemical mutagenesis (Auerbach and Westergaard, 1960). Of course this was before the definition of mutation in terms of base changes and so on. But he covered the different classifications from a biological point of view very well.

Auerbach: He only dealt with what he called specificity "at the geographical level".

Loveless: He had about five categories.

Auerbach: Yes, but the specificity which arises later in the mutation process was not included in any of these.

Kubitschek: At the chemical level, depending on what model of replication of DNA one prefers, one would have to distinguish between mutagen specificity and strand specificity. On a master strand model one would predict two different types of "repair" or kinetics of mutation. If repair were absent one would predict that a master strand would give rise to mutations ultimately in all progeny, while a mutated complementary strand must be diluted out as cell numbers increase, with an exponential drop-off of 50 per cent at every cell division.

Auerbach: This is quite a different type of specificity.

Kubitschek: Yes.

Apirion: Maybe we should call one primary and the other secondary specificity. In the end a mutation is a specific change in DNA and it doesn't matter how it comes about. But in one case the mutation is caused by a direct reaction of the mutagen with DNA, while in the other case the mutation is caused by an interaction between DNA and a product that was produced by the reaction of the mutagen with some other cellular component.

Auerbach: I used to apply the terms "primary" and "secondary" but I didn't like them because I didn't want to imply that secondary specificities are of secondary importance.

REFERENCES

Allison, M. (1969). *Mutation Res.*, **7**, 141–154.

Auerbach, C., and Westergaard, M. (1960). *Abh. dt. Akad. Wiss. Berl.*, Kl. Med. No. 1, 116–123.

Clarke, C. H. (1962). *Z. VererbLehre*, **93**, 435.

Kølmark, H. G., and Kilbey, B. J. (1962). *Z. VererbLehre*, **93**, 356–365.

Kølmark, H. G., and Kilbey, B. J. (1968). *Molec. gen. Genet.*, **101**, 185.

Yanofsky, C. (1968). *XII Int. Congr. Genet.*, Tokyo.

REPAIR OF LATENT T5-RESISTANT MUTANTS IN CHEMOSTAT CULTURES

H. E. KUBITSCHEK

with the technical assistance of H. E. Bendigkeit*

Division of Biological and Medical Research, Argonne National Laboratory, Argonne, Illinois

THE linear accumulation of T5-resistant mutants in chemostat cultures of *Escherichia coli*, first demonstrated by Novick and Szilard (1950), provided evidence that these phenotypic mutants are not under selection. We confirmed this by reconstruction experiments (Kubitschek and Bendigkeit, 1964*b*). In addition, selection for *latent mutants*—cells genetically mutant or in which the processes leading to mutation have been initiated, but in which the mutant character is not yet expressed—was apparently absent.

Two types of experiments provide evidence against selection of latent T5-resistant mutants. In the first, when cultures were continuously exposed to a mutagen at relatively low concentrations or dose rates, latent mutants were produced and phenotypic mutants were expressed at the same constant rate. This result was obtained with caffeine, ultraviolet light (u.v.), and 2-aminopurine (Kubitschek and Bendigkeit, 1961, 1964*b*), with acridine orange/light (Webb and Kubitschek, 1963), and with ^{60}Co γ-rays (unpublished results). In the second type, when cultures were exposed only briefly to mutagens at higher concentrations or dose rates, initial frequencies of latent mutants produced were equal to the frequencies finally expressed. This result was obtained with acridine orange/light, methylene blue/light, and 2-aminopurine (Kubitschek, 1964, 1966). Thus, no repair or selection was observed for T5-resistant latent or expressed mutants in any of these chemostat cultures.

This absence of selection is of special interest because of its implications regarding the nature of replication of deoxyribonucleic acid (DNA). According to the conventional Watson-Crick model of replication, any mutagen that affects only a single strand of the DNA duplex must give rise

* Present address: Department of Biological Science, Stanford University, Stanford, California.

to latent mutant frequencies that decrease at the first division after mutation because of segregation of parental strands to the two daughter cells. Yet there was no evidence for segregation of the mutational lesion with any of the mutagens mentioned, although some or all of them are expected to act upon single strands, either partly or exclusively. In another mutational system, mutation from tryptophan requirement to prototrophy, Bridges and Munson concluded that nuclear segregation occurred, but that *strand* segregation was absent after mutation. They found no strand segregation after mutations were induced in steady-state cultures with γ-rays, X-rays, and u.v. (Bridges and Munson, 1964*a*, *b*, 1966, 1968*a*, *b*; Munson and Bridges, 1964).

In contrast to these steady-state results, when chemostat cultures were exposed briefly to u.v. at doses large enough to kill 25 per cent or more of the cells latent mutants were often lost from the population. In addition, final expressed mutant yields depended upon generation times in these cultures. These results suggested that u.v.-induced mutational lesions might be repaired under some conditions in chemostat cultures, and so led to the experiments reported here.

METHODS

Cultures

Chemostat cultures of *Escherichia coli*, strains B, B/1, *try*$^-$, and B/r/1, *try* were grown at 37°C in the M9-salts solution previously described (Kubitschek and Bendigkeit, 1964*a*), in the presence of a limiting concentration of glucose (100 μg/ml). L-Tryptophan was added to cultures requiring tryptophan, to give a concentration of 5 μg/ml (approximately a 10-fold excess over the limiting concentration).

Assays for mutant frequency

Concentrations of viable cells and mutants resistant to bacteriophage T5 were determined by standard methods, described earlier in detail (Kubitschek and Bendigkeit, 1961). Briefly, phenotypically *expressed mutants* were detected by spreading chemostat samples upon nutrient agar or salts-lactate-glucose plates, which were then immediately sprayed with T5. Expressed mutant frequencies (*E*) were calculated from colony counts observed on these plates. *Latent mutants* (*L*), cells phenotypically wild but destined to produce mutant progeny because of exposure to mutagen, were detected by growing them on nutrient agar plates for 3 hours or salts-lactate-glucose for 4·5 hours at 37° (about five or six generations) before spraying with T5. Almost all latent mutants gave rise to one or more expressed mutants among their progeny during this period, permitting the development of a resistant colony. Since expressed mutants also produce resistant colonies on these plates, these colony counts provided a measure of the sum $E+L$ of expressed plus latent mutant frequencies. Mutants arising spontaneously upon these plates appeared in negligible frequencies when nutrient agar plates were sprayed before 3·5 hours or salts-lactate-glucose plates were sprayed before 5·5 hours.

U.v. irradiations

Cultures (250 ml, 3×10^8 cells/ml) were irradiated at 37° in chemostat growth tubes through a quartz port 3 cm in diameter. The u.v. source was a 2·5-cm section of a germicidal lamp (General Electric, 15 W) placed at a distance of 80 to 120 cm from the port. Average dose rates were about 0·05 erg/mm^2 per second. Although nutrient flow to the culture was blocked for half an hour before irradiation, aeration was continuous. In all experiments, at least 25 per cent of the cells survived irradiation. After the irradiation, nutrient flow was blocked for a further period of 3 to 18 hours, with the temperature maintained at 37°. At the end of this time, part of the culture was used immediately, with the remainder refrigerated at 4° for further use. For measurements of mutation kinetics, two or more smaller chemostats (23 to 25 ml) were filled with the irradiated culture.

SELECTION AGAINST LATENT MUTANTS

After irradiation at u.v. doses sufficient to kill more than 25 per cent of the cells, measured values of expressed E and latent L mutant frequencies remained essentially unchanged during the initial stationary period before nutrients were again supplied (Figs. 1A and 2). After nutrient flow was re-established, values of L increased rapidly and reached a maximum value shortly after the time required for a single generation during steady-state growth. Since these latent mutants must have been created by exposure to u.v., and therefore must have been present from the time of irradiation, it was clear that this initial increase in L is an artifact arising in the test for latent mutants.

This conclusion was confirmed by an examination of the time required for expression of latent mutants in samples taken from chemostats at successive intervals after irradiation. As shown in Fig. 3, for one such experiment, the earliest samples have an extended lag period before phenotypic mutants are produced on nutrient agar plates. Later samples

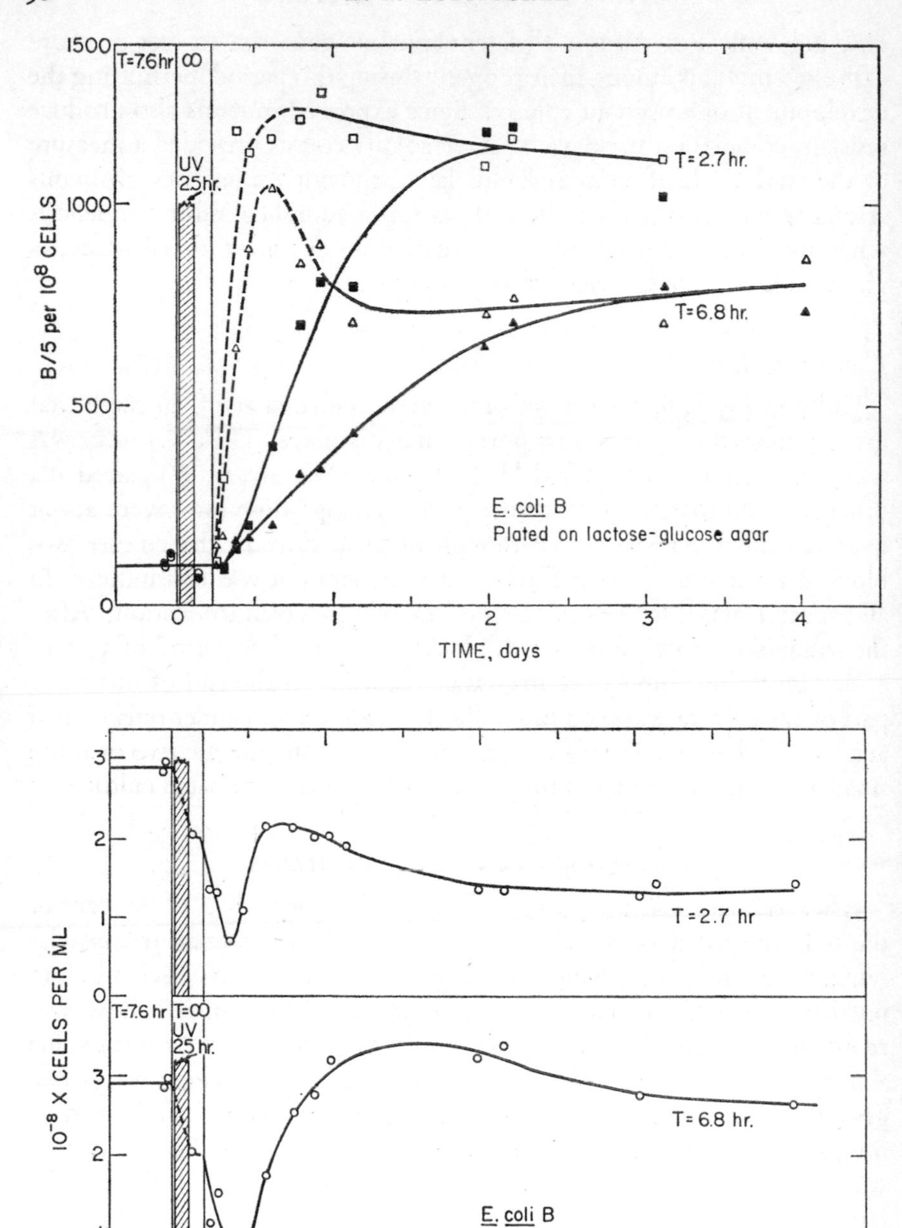

Fig. 1. (B) Cell concentrations in the chemostats used for (A).

FIG. 1. (A) Mutant B/5 frequencies at two wash-out rates after u.v. irradiation as measured by plating on salts-lactate-glucose agar. (Redrawn from Fig. 8, Kubitschek and Bendigkeit, 1961). *E. coli* B at 37°C in chemostats limited with glucose, 150 μg/ml. Cell survival after irradiation: 69 per cent. The irradiated culture was used to fill the growth tubes of two smaller chemostats with wash-out times corresponding to steady-state cell generation times of 2·7 and 6·8 hours. □, ▲; expressed mutant frequencies (E). □, △: sums of expressed plus latent mutant frequencies ($E+L$).

have successively shorter lag periods, reaching a constant minimum value after about the first generation time. After this time, almost every latent mutant produces one or more expressed mutant progeny within about

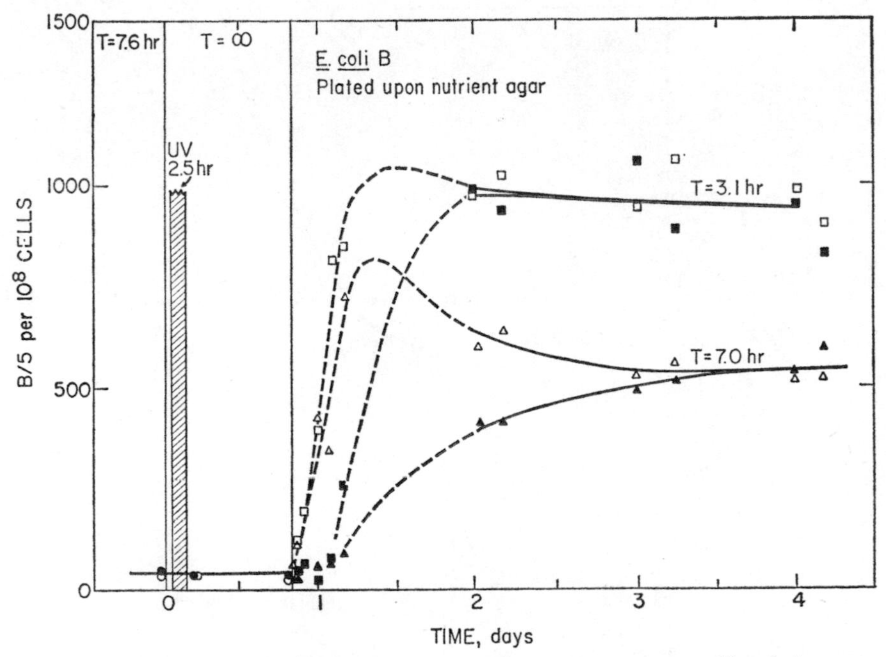

FIG. 2. U.v.-induced mutant frequencies at two wash-out rates as measured by plating upon nutrient agar. Cell survival after irradiation: 59 per cent. Same culture conditions and symbols as for Fig. 1.

2·5 hours at 37° on nutrient agar. Parallel but longer lag periods were observed when cells were plated on the minimal salts-lactate-glucose agar used in the experiment in Fig. 1.

It may be that other undetected mutations are produced and repaired during the first equivalent chemostat generation time, but if so, their disappearance makes them unsuitable for our investigation. Evidence will be presented that the latent mutants that remain are due to unrepaired lesions.

At the most rapid growth rates, such as those in Figs. 1A and 2, decreases from the early maximum value of the sum $E+L$ were barely perceptible or could not be distinguished from values at later times, or from the final mutant yield E_∞ after all latent mutants were expressed. At slow growth rates, on the other hand, mutant yields decreased markedly from the maximum value $(E+L)_{max}$

$$E_\infty < (E+L)_{max} \tag{1}$$

These decreases indicated that some latent mutants failed to give rise to mutant progeny in these chemostat cultures. This is selection against latent

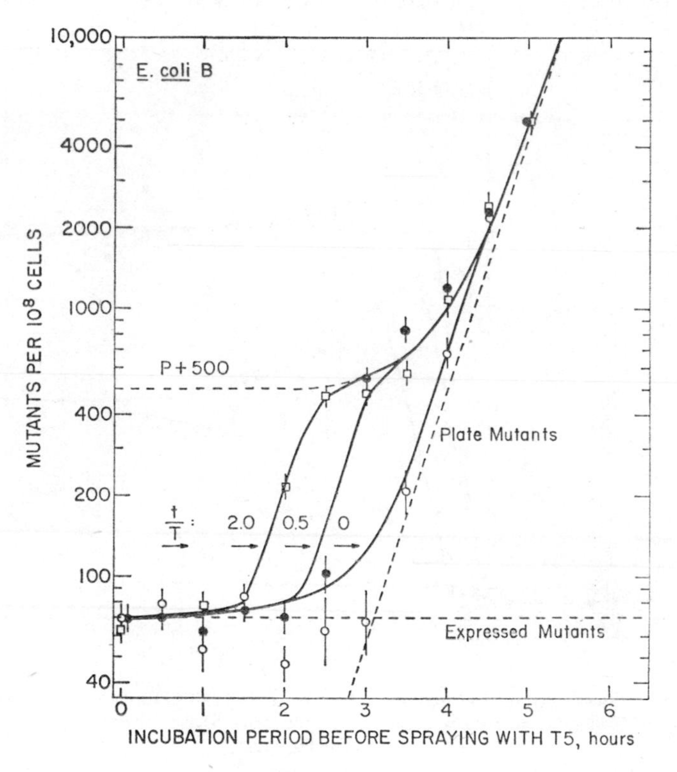

FIG. 3. Effect of period of incubation upon the appearance of T5-resistant colonies on nutrient agar plates for samples collected at three different chemostat ages after u.v. irradiation. Samples were collected from the chemostat at times equivalent to 0, 0·5, and 2·0 elapsed generation times t/T during steady-state growth. Cells were spread upon nutrient agar plates and incubated at 37°C before being sprayed with T5. Dashed lines show frequencies of expressed mutants E and plate mutants P. Plate mutants were assumed to double at the rate of increase in cell number, a 20-min doubling time. The uppermost curve, extended by a dashed-line curve ($P+500$), represents the maximum frequency that would be observed if all latent mutants were expressed. ○: 0 generations; ●: 0·5 generations; □: 2·0 generations.

mutants in the chemostat, and could occur through either their reduced growth rates, their death, segregation of some single-stranded mutational lesions, or repair of mutation.

RATES OF DISAPPEARANCE OF LATENT MUTANTS

Latent mutant frequencies were calculated from data of the kind shown in Figs. 1 and 2 by subtracting values for the frequencies E of expressed mutants from those for the sum $E+L$. The rate of decline of latent mutant frequencies was exponential:

$$L=L_0e^{-t/D} \qquad (2)$$

where L_0 can be any arbitrary initial value after the first generation, L is the latent mutant frequency at the later time, t, and D is a constant.

The value of D appeared to be independent of the growth rate of the culture or the strain used, whether B, B/1, *try*⁻, or B/r/1, *try*. Fig. 4 gives an example of the decay of latent mutant frequencies, showing measured values of L when an irradiated culture of *E. coli* B was partitioned into six

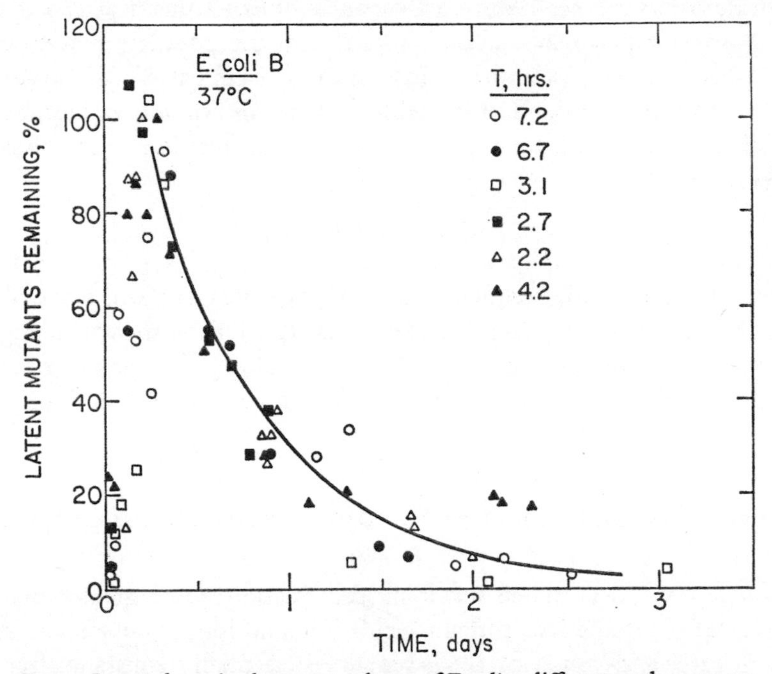

FIG. 4. Latent decay in chemostat cultures of *E. coli* at different wash-out rates. All cultures were taken from the same irradiated chemostat culture. Latent mutant frequencies are shown as a percentage of the maximum value. The initial increase to the maximum value is artificial, as explained in the text.

daughter chemostats with different generation times, from 2·2 to 7·2 hours, all at 37°. Latent mutant frequencies decayed at essentially the same exponential rate in all of these cultures, with a half-value period of approximately 11 hours.

Within the limits of experimental error, the same half-value was also obtained for stationary cultures maintained at 37° without the addition of a carbon source after collection. The mean half-value for decay at 37°, calculated from 20 experiments with both growing and stationary cultures for each of which six to 20 values of L were determined, was 10·55 hours, with a standard error of 0·16 hour. During the first half-value period or two in stationary cultures there was little or no observable cell death.

TEMPERATURE DEPENDENCE OF LATENT MUTANT DECAY

Rates of decay of latent mutant frequencies were also measured in stationary cultures at 5, 18, 25, and 44°. Average half-values (D ln 2) at these temperatures were 1340, 660, 68, and 1·7 hours, respectively. There was little or no cell death during the course of experiments at 18 and 25°. In the other experiments, about one-third of the population were dead after 3 hours at 44°, 48 hours at 37°, and two weeks at 5°. These results, along with those obtained at 37° with growing and stationary cultures, fit an Arrhenius equation for the rate of decay as a function of the absolute temperature:

$$k = Ae^{-E/RT} \tag{3}$$

where A is the reaction frequency, E the energy of activation, R the molar gas constant, and T the absolute temperature. This is shown in Fig. 5, where k is plotted as the reciprocal of the half-value. The energy of activation calculated from these data is $E = 25$ kcal/mole, and $A \approx 10^{12}$/sec.

PREVENTION OF MUTATIONAL LOSS: COMPENSATED CHEMOSTATS OR LOW DOSES

At low u.v. doses, when surviving cell fractions were greater than 90 per cent after irradiation, mutant yields were no longer dependent upon growth rate, and values of $E+L$ remained essentially constant after the first chemostat generation:

$$E_{\infty} = (E+L)_{max} \tag{4}$$

In these cultures, cell concentrations did not deviate greatly from their steady-state values, since the rate of supply of glucose was approximately equal to the rate of utilization. In contrast, at high doses, when cell survival was about 70 per cent or less, viable cell concentrations first decreased by a factor of two or more and the rate of utilization of glucose was reduced, causing glucose to accumulate in the growth tube. During this period,

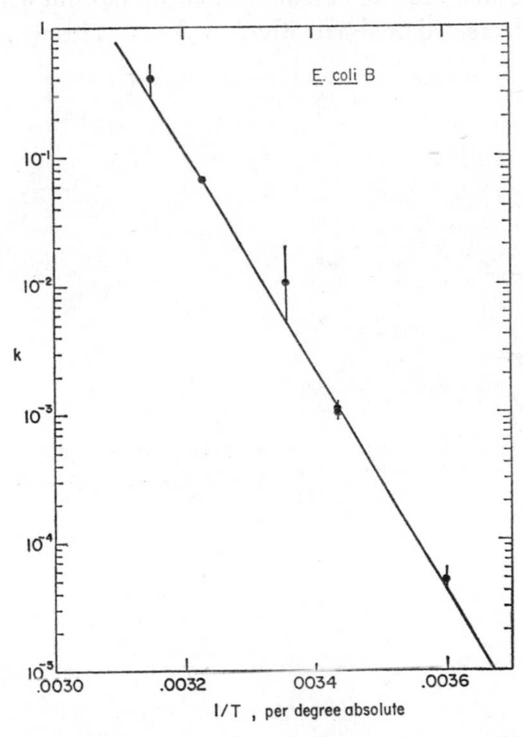

Fig. 5. Arrhenius plot for latent mutant decay rates in growing and in stationary cultures. Values of the ordinate are shown as $k=1/(D \ln 2)$, the reciprocal of the half-value for decay. These values are averages obtained with all three strains of *E. coli*.

cells were washed out, apparently because there was little or no division. When division began again cell concentrations increased temporarily to values beyond the steady-state concentration.

Because these large departures from steady-state conditions might be associated with onset of the loss of latent mutants, we examined the kinetics of mutation after high doses of u.v. in "compensated" chemostats. In these chemostats, we reduced the concentration of glucose supplied to the culture to match the level of surviving cells, i.e. both were reduced by the same fraction. The results of these experiments, shown in Fig. 6, indicate that mutant yields in compensated chemostats are independent of growth rate, and that all latent mutants ultimately give rise to completely mutant progeny—expression is complete.

Since there was no evidence for the loss of latent mutants in growing cultures after low doses of u.v. (surviving fractions greater than 90 per cent), it seemed that the rate of loss of latent mutants in stationary cultures might also be decreased or absent after low doses. This was indeed the case,

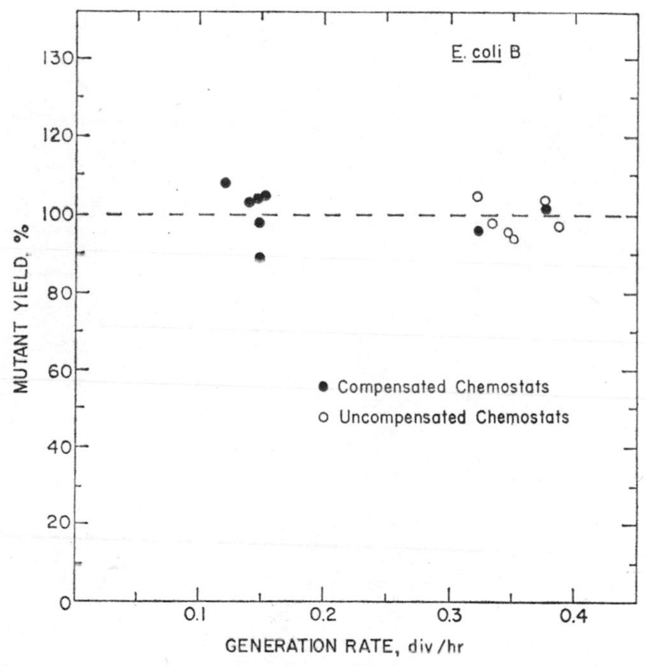

FIG. 6. Mutant yields in compensated chemostat cultures. Yields (E_∞) are given as a percentage of the yield observed in very rapidly growing uncompensated cultures after exposure to high doses of u.v.

as shown in Fig. 7, which compares the loss of latent mutants in two stationary cultures at 44°, established after one chemostat generation. One of these was exposed to a relatively large dose of u.v. (73 per cent survival) and the other to a low dose (no killing was observed). The rate of decay was much greater in the culture exposed to the higher dose of u.v.

EVIDENCE FOR REPAIR OF MUTATIONAL LESIONS

In the experiments reported here, there was little or no selection of u.v.-induced mutants in steady-state cultures, but when growth conditions were disrupted by appreciable killing, latent mutant frequencies decreased markedly at slow growth rates. Because of their low frequencies, it was not possible to isolate latent mutants and make direct observations that would distinguish between the alternative possibilities for this selection: mutant death, decreased growth rate, or segregation or repair of mutational lesions. There is, nevertheless, good evidence to support repair.

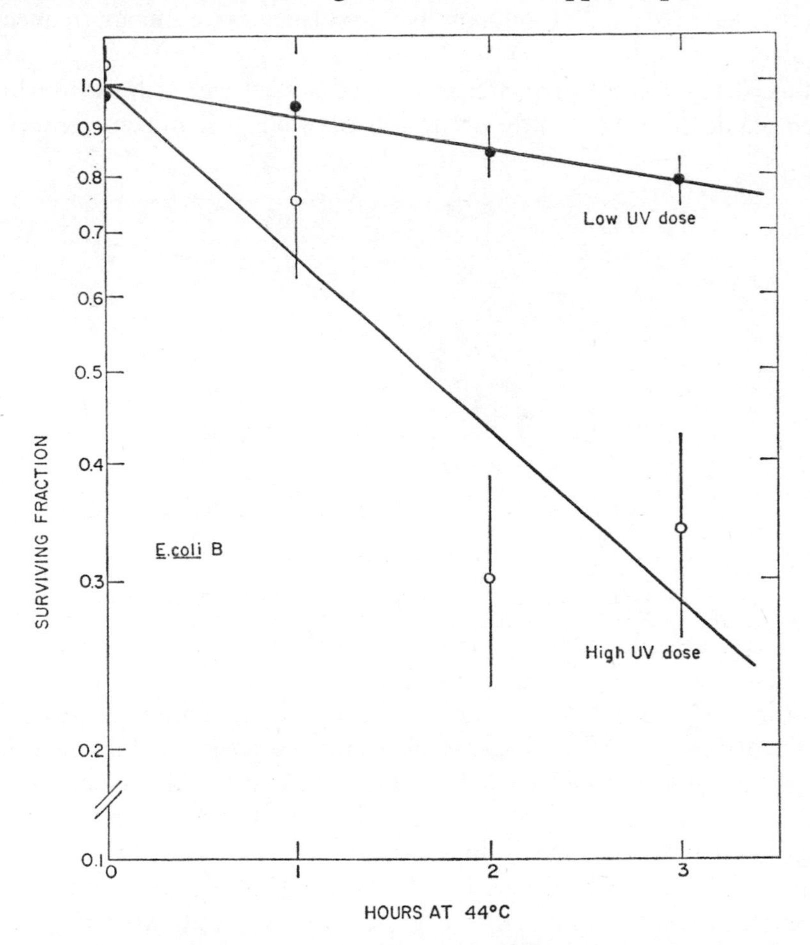

FIG. 7. Effect of high and low u.v. dose upon the decay of latent mutants in stationary cultures at 44°C. Cell survival after 3 hours was approximately 85 per cent after the high dose and 95 per cent after the low dose of u.v.

The formation of filamentous cells after irradiation with u.v., known to occur in the B strains, would have reduced cell division rates of mutants. We searched for filamentous cells in these irradiated cultures but, with the exception of an aberrant culture or two, failed to find them. Furthermore, mutant frequencies also decreased at the same rate in the filament-resistant strain, B/r/1, *try*. More important, mutant frequencies also decreased in stationary cultures in the absence of an energy source for growth and division, and filamentation *per se* cannot account for this decrease. This decay of latent mutants in stationary cultures, together with the fact that earlier studies provided no evidence for a change in growth rate, makes a decreased rate of cell division an unlikely explanation for mutant frequency decline.

Since latent mutant frequencies decreased in stationary cultures in which there could have been little or no cell division, this mutant frequency

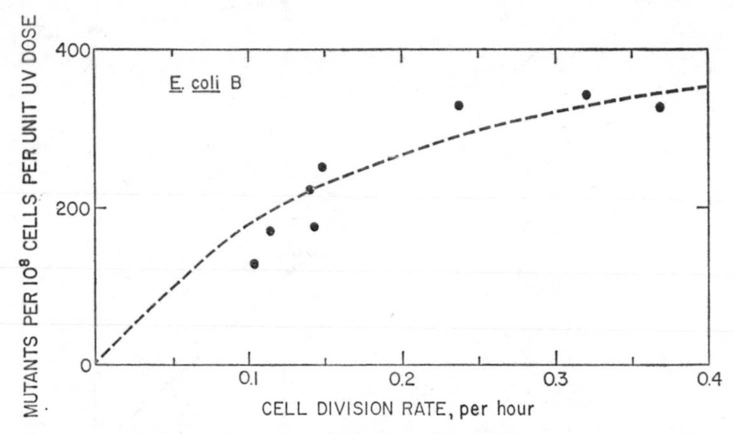

FIG. 8. Dependence of mutant yield upon cell division rate in chemostats. (Redrawn from Fig. 9, Kubitschek and Bendigkeit, 1961, and corrected.) The dashed line shows the values predicted by Equation 5 for a value of $C=0{\cdot}5$.

decline could not have been due to a segregational division in these cells. In further support of this conclusion, a simple segregational division, had it been possible, should have led to a mutant frequency decline of only 50 per cent, rather than the very low values that actually were observed, which were less than 10 per cent in some cases.

If latent mutants died because of associated lethal lesions, one might expect cell death by or before cell division. Yet cells were frequently spared for several generations after irradiation, only to die afterwards. Again, if such lesions existed, they must also have been produced in irradiated cultures grown in compensated chemostats. But those mutant yields were not affected. It would be necessary, then, to postulate a repair mechanism for these lesions that would otherwise have proved lethal. Although such complexities cannot be ruled out at present, it is simpler and more direct to postulate instead the repair of mutational lesions.

The repair of mutational lesions is consistent with all the observations already reported, as well as with other findings to be reported later. First, all latent mutants must be *genetic* mutants before the end of the first division in the chemostat, since the mutational change was transmitted to all progeny in compensated chemostats. Second, if repair occurs, the rate of repair of latent mutants in stationary cultures follows an Arrhenius relationship, as expected for an enzymic system capable of repair. Third, the repair system was not activated in cultures in which glucose concentrations remained limiting. This result is analogous to the finding by Setlow and Carrier (1964) that excision of dimers from DNA in related strains of *E. coli* is reduced or inhibited by starvation for glucose. Together these results suggest that starvation for glucose prevents production or activation of the excision enzyme. Since glucose was not supplied when repair took place in stationary cultures, the energy must have been made available by endogenous metabolism.

PHOTOREACTIVATION

The above results suggested that these latent mutants might be repaired by exposure to visible light. Latent mutants were photoreversed by more than 60 per cent after an exposure to visible light (approximately 10^6 erg/mm^2 in 2 hours) at 26° that caused little or no cell death.

COMPETITION BETWEEN REPAIR AND EXPRESSION

Earlier results obtained for mutant yields as a function of generation time (Kubitschek and Bendigkeit, 1961) and replotted in Fig. 8 can be fitted by assuming that repair processes compete with expression of latent mutants. The fit is approximate, since we neglect any possible decrease in concentration of repair enzymes, such as might occur during growth and dilution during the return to steady-state conditions. We assume that latent mutant decay in the absence of cell division represents complete repair, since there is no opportunity for cell division and expression of the mutation. Then, from Equation (2), the rate of repair ρ_r is given by

$$\rho_r = 1/D$$

When cells divide, some latent mutants give rise to phenotypic mutants. The rate of expression, ρ_e, should be proportional to the rate of cell division, and, therefore, to the reciprocal of the generation time, T:

$$\rho_e = C/T \quad (C = \text{constant})$$

Then, the mutant yield is proportional to the rate of expression:

$$\frac{E_\infty}{E_{max}} = \frac{\rho_e}{\rho_e + \rho_r} = \frac{CD}{CD + T}, \tag{5}$$

where E_{max} is the mutant yield when all latent mutants are expressed. Equation (5) fits the data of Fig. 8 well when C has a value of 0·5, as shown.

With this value of C, the rate of expression of a latent mutant is 0·5; that is, on the average, division of a latent mutant gives rise to one latent mutant daughter and one expressing the mutation, after correcting for repair. This result is the same as that found for the decrease in latent mutant frequency when mutations were induced by acridine orange/light in a B/r strain (Kubitschek, 1964). There is no apparent decrease in mutant frequency as a result of a segregational division in either system.

ACTIVATION OF THE REPAIR SYSTEM

In summary, the evidence we have discussed supports the following conclusions for mutation to T5-resistance in chemostat cultures:

(1) There is no detectable repair of latent mutants for T5-resistance in steady-state chemostat cultures, whether these mutations are induced with u.v. or with other mutagens.

(2) When more than 25 per cent of the cells are killed by exposure of chemostat cultures to u.v., the growth rate of the remaining culture is first decreased and then increased above steady-state levels in the absence of compensatory measures. Under these conditions, repair occurs in the absence of visible light in growing or stationary cultures. More than 90 per cent of the latent mutants are reparable in stationary cultures. Expressed mutants are not repaired. The rate of repair in the dark follows an Arrhenius relationship, with an energy of activation of about 25 kcal/mole.

(3) Exposure of latent mutants to visible light permits repair of more than 60 per cent of them, with little or no observable cell death.

(4) There is no evidence for loss of newly-induced mutants for T5-resistance during a segregational division, whether repair occurred or not.

The first three conclusions are consistent with the interpretation that repair enzymes affecting mutation to T5-resistance are not normally produced under the severely growth-limiting conditions of steady-state chemostat cultures, but that activation or synthesis of these enzymes occurs during the loss of steady-state conditions after u.v. doses large enough to kill more than 25 per cent of the cells. As the culture returns to steady-state conditions, repair enzymes may no longer be produced, but would be expected to be diluted out to daughter cells. Any repair that occurs after cell concentrations return to normal values would then be attributed to the continued function of those enzymes.

The fourth conclusion, taken together with the absence of segregational loss of mutants with the many different mutagens that have been used, is unexplainable by the conventional Watson-Crick mechanism for DNA replication.

REFERENCES

Bridges, B. A., and Munson, R. J. (1964*a*). *J. molec. Biol.*, **8**, 768–770.
Bridges, B. A., and Munson, R. J. (1964*b*). *Mutation Res.*, **1**, 362–372.
Bridges, B. A., and Munson, R. J. (1966). *Biochem. biophys. Res. Commun.*, **22**, 268–273.
Bridges, B. A., and Munson, R. J. (1968*a*). In *Current Topics in Radiation Research*, pp. 95–188, ed. Ebert, E. and Howard, A. New York: Wiley.
Bridges, B. A., and Munson, R. J. (1968*b*). *Proc. R. Soc. B*, **171**, 213–226.
Kubitschek, H. E. (1964). *Proc. natn. Acad. Sci. U.S.A.*, **52**, 1374–1381.
Kubitschek, H. E. (1966). *Proc. natn. Acad. Sci. U.S.A.*, **55**, 269–274.
Kubitschek, H. E., and Bendigkeit, H. E. (1961). *Genetics, Princeton*, **45**, 105–122.
Kubitschek, H. E., and Bendigkeit, H. E. (1964*a*). *Mutation Res.*, **1**, 113–120.
Kubitschek, H. E., and Bendigkeit, H. E. (1964*b*). *Mutation Res.*, **1**, 209–218.
Munson, R. J., and Bridges, B. A. (1964). *Nature, Lond.*, **203**, 270–272.
Novick, A., and Szilard, L. (1950). *Proc. natn. Acad. Sci. U.S.A.*, **36**, 708–719.
Setlow, R. B., and Carrier, W. L. (1964). *Proc. natn. Acad. Sci. U.S.A.*, **51**, 226–231.
Webb, R. B., and Kubitschek, H. E. (1963). *Biochem. biophys. Res. Commun.*, **13**, 90–94.

DISCUSSION

Kimball: In the test for latent mutant expression do you wait a while before making the mutation test, whereas when you don't look for latent mutants do you test for the presence of the mutation immediately?

Kubitschek: We transfer culture samples to two sets of plates and do parallel tests on these plates at two different times.

Apirion: Most of these studies were done, I presume, before some of the strains were available. If most of the loss of latent mutation is due to repair of damaged DNA it would perhaps be useful to repeat some of the experiments with *hcr*⁻ strains, for example.

Kubitschek: I did many of the pulse experiments with the repair-resistant strains, using the mutagens acridine orange and light, methylene blue and light, and 2-aminopurines. I found essentially the same kind of kinetics I showed you for the experiments in which cell death was avoided. Under those conditions there is no evidence for any repair or for any molecular segregation; if a single strand is mutated the mutation is somehow always transmitted to all the progeny of the cell.

Apirion: Do other mutagens show the one-strand effect?

Kubitschek: Yes, possibly acridine orange/light and methylene blue/light.

Bridges: It has always worried me that in your cultures you didn't get any segregation at all. All the experiments were done in very slowly growing cultures. Do you also get no segregation if you raise the growth rate as high as possible—to generation times of the order of, say, one hour?

Kubitschek: I have never grown chemostat cultures so rapidly. It wouldn't be a good chemostat, stably controlled, in those circumstances.

Bridges: How many nuclei do these cells have? I imagine they have only one visible one.

Kubitschek: I think that they must have a single replicating nucleus. Maaløe and others have shown that the content of DNA per cell goes down to reach a minimum amount in very slowly growing cultures (Schaechter, Maaløe and Kjeldgaard, 1958; Herbert, 1959). The amount of DNA per cell changes at some point between a half and one division per hour, as R. Ecker and M. Schaechter (personal communication) found with *Salmonella*, for example.

Bridges: So it is nearly all G1, if you like that terminology.

Maaløe: I don't think we need to quote anybody to know how many nuclei slow-growing cells have. At the time of cell division the minimum is two complete genomes. Those two have to be created from one. Unless there is a considerable gap between successive rounds of replication early in the division cycle, a site close to the origin of replication may be present in two copies most of the time even in uninucleated cells.

You mentioned that you think the cultures grew with an effective glucose concentration of about 1 μg/l. Did you check this in the same way Novick and Szilard (1950) tested tryptophan, by determining growth rates in extremely dilute cultures at low, fixed concentrations of the test compound?

Kubitschek: No, this estimate of residual glucose concentration is made on the basis of the linearity that one observes at different glucose concentrations in the chemostat. So I might be off by a factor of 2 or 3, but not much more than that. The point I was making about the DNA per cell was, as in Ecker and Schaechter's experiments, that a plot of the amount of DNA per cell did have a horizontal slope at low growth rates. This means that in the region where we generally operate chemostats, one would expect to find a minimal amount of DNA. When we studied the uptake of tritiated thymine into DNA, we found a long G1 period in *E. coli* cultures that grew with a doubling time of two hours or longer: about two-thirds of the cycle had passed before DNA synthesis began.

Dawson: When you plate the cells do you always find that there is a lag phase before the cells enter the exponential phase on the plates?

Kubitschek: Yes. This is quite a change of growth conditions.

Dawson: You spray after three hours, and the relatively large number of mutants which you then score, minus the number you scored earlier, you call latent mutants. Presumably if they are latent mutants that have undergone some divisions, and are becoming expressed, they should be appearing in clones. What is the evidence that the excess mutants that you get after three hours are appearing in clones?

Kubitschek: We count colonies on plates and spray at successive time intervals to measure the mutant frequency. The mutant frequency starts off at some level due to expressed mutants and then, as the cells grow and multiply, they go through a second period in which the latent mutants which were present have now had an opportunity to give rise to progeny which also express mutation.

Dawson: Is it necessary to talk about these as latent mutants and ask questions about segregation and so on? Might not the cells have physiological changes induced by the u.v. treatment, such that there is a higher general rate of mutation at the transition of lag to log on the plates? It is possible to interpret the data that Luria and Delbrück (1943) and Newcombe (1948) obtained from experiments using many small cultures in terms of considerable peaks of mutation at transition from lag to log and from log to resting.

Kubitschek: The best evidence for constant mutation rates is internal. When a mutagen is added to a culture, then the rate of increase of mutants will depend upon the number of mutants already there and the rate of cell division; it will also depend upon the mutation rate per cell and the number of mutants in the population. If there is no selection the rate of cell division in these cultures will match the rate of wash-out, and the rate of mutation will increase linearly. This was Novick and Szilard's (1950) argument. If instead there were some transient effect then one would see a bump in the rate of accumulation of mutants.

Maaløe: Aren't your mutants mostly cells that don't absorb phage T5? If so, there is an unknown but fairly long interval, *after* nuclear segregation, before the cell surface no longer offers an appreciable number of phage receptor sites. Has this time interval been estimated in your experiments?

Kubitschek: This model (of complete expression of mutation) is in agreement with the kinetics of expression of mutation in a culture exposed to acridine orange and light. It assumes that in this particular case one of the strands, which is a master strand, has the mutation on it. The master strand model is different from the Watson-Crick model in that the master strand must code for the new complementary strand in one of the cells and also for the new master strand in the other. In that way we would get the same kind of segregation of expression of mutation in one daughter cell, but the other daughter cell would be a double-stranded mutant, and would express mutation. With this kind of kinetics, where a latent mutant gives rise to both an expressed mutant and another latent mutant of the same form, the mean delay period should be 2·47 cell divisions, provided that we allow for the observation that no mutants are expressed until one cell division has elapsed.

It was earlier suggested that in some cells presumably part of the DNA is replicated, and in these cells, if there is only mutation of one strand, then we should see a segregational division according to any model. Since there is no experimental evidence for a segregational division this is a mystery. The only thing I can say is that on a master strand hypothesis this kind of mutation would segregate out after one subsequent division; on the Watson-Crick model it would require two subsequent cell divisions, so on this latter model the absence of segregation is an even bigger mystery.

Evans: Did you say that under the conditions of low growth rate something like two-thirds of the cells were in a so-called G1 state?

Kubitschek: In the experiments that we have done, at slow growth rates in chemostat cultures, we found a G1 period which progressively lengthened to two-thirds of the cycle (Kubitschek, Bendigkeit and Loken, 1967). At a generation time of one hour DNA synthesis began immediately in newly divided cells. At generation times of two hours and longer, DNA synthesis was delayed for two-thirds of the cycle. There is also some evidence for a G2 period in addition to a G1, but there are no data on the length of the G2 period.

Evans: The question that I really want to put is, is the change in repair capacity in the two different culture states a consequence of a difference in the interphase structure of the cell populations in these two conditions? For instance, is it possible that the increased availability of a carbon source is not itself directly responsible for allowing repair to occur, but in fact simply allows the cells to proceed from an interphase stage—perhaps in G2—in which little repair can take place, into a stage—perhaps S—where repair processes may be much more active?

Kubitschek: A large fraction, perhaps 25 per cent, of cells are in the S phase, so we should see some repair, but we see none.

Evans: Have you in fact examined the structure of the interphase population as between the two kinds of culture states and if so do you observe changes?

Kubitschek: We haven't done any biochemical tests. There may have been a few filaments in an aberrant culture or two, but one usually cannot see filaments and also the B/r strains are filament-resistant. I think that the repair enzyme system is activated in that early portion of growth after u.v. irradiation where the population at first drops in concentration, and then rises again. I don't think activation is due to increased glucose concentration alone. We tried to test this by inducing mutants at low frequency so that the repair system wouldn't be activated, and then we artificially increased glucose concentrations, but we were never able to demonstrate activation of the repair system under those circumstances.

Apirion: The repair mechanism is apparently very important. Is there an *in vitro* assay by which one could follow the amount of repair activity at various stages of the growth cycle, and at various growth rates?

Kimball: One can certainly follow the loss of thymine dimers and methods are now becoming available with which one can follow repair synthesis, or non-conservative replication.

Kubitschek: I think it is unlikely that these mutations are caused by thymine dimers. In some preliminary experiments we killed cells from chemostat cultures and examined repair in these. Their kinetics of repair in the same conditions are at least four times faster than repair of mutational lesions. Instead of two hours, it took about 15 minutes to photoreactivate the cells at the same light intensity. That is, photoreactivation of cells which would otherwise have died occurred with much more rapid kinetics than photoreversal of mutations.

Apirion: You mentioned that a strain carrying an hcr^{-} mutation behaved the same way as an hcr^{+} strain. However, since complete loss of the HCR system could be lethal, such strains could be only partial mutants.

Kubitschek: As I said before, we do not know what was happening in the first generation. Many mutants may have been repaired, perhaps by other processes. In these experiments all we could study was the repair of mutants which survived after the maximum mutant production was reached.

Auerbach: Repair has become highly fashionable now, but one should be very cautious in assuming that a phenomenon is due to repair when there is absolutely no evidence for it. The danger about this assumption is that it bars the way to other means of exploration. In your case, killing rather than repair might be responsible.

Kubitschek: I agree with you wholeheartedly. At the risk of putting my neck out even further I shall tell you about some experiments which we haven't had a chance to repeat yet. With u.v. in the B/r hcr^+ strains, the production of latent mutants seemed to give rise to full mutant yields. In the hcr^- strain, where repair didn't seem to occur, all the mutants decayed at the rate of 50 per cent of the latent mutant concentration with every successive cell division.

Auerbach: So they are repaired in the hcr^-, but not in the hcr^+?

Kubitschek: On the face of it, but one can make other interpretations.

REFERENCES

HERBERT, D. (1959). In *Recent Progress in Microbiology*, pp. 381–396, ed. Tuneval, G. Springfield, Ill.: Thomas.

KUBITSCHEK, H. E., BENDIGKEIT, H. E., and LOKEN, M. R. (1967). *Proc. natn. Acad. Sci. U.S.A.*, **57**, 1611.

LURIA, S. E., and DELBRÜCK, M. (1943). *Genetics, Princeton*, **28**, 491–511.

NEWCOMBE, H. B. (1948). *Genetics, Princeton*, **33**, 447–476.

NOVICK, A., and SZILARD, L. (1950). *Proc. natn. Acad. Sci. U.S.A.*, **36**, 708.

SCHAECHTER, M., MAALØE, O., and KJELDGAARD, N. O. (1958). *J. gen. Microbiol.*, **19**, 592.

GENERAL DISCUSSION

Devoret: In relation to the problem of the influence of the cell on the expression of mutations, may I report briefly some experiments based upon a phenomenon discovered by F. Jacob in 1954?

Jacob demonstrated that if *Escherichia coli* K12 bacteria were u.v.-irradiated and then infected by intact—non-u.v.-irradiated—phage λ11 particles, the phage mutation rate to virulence was increased by a 30- to 40-fold factor over the background. As the capacity of the cells to reproduce the phage was not affected in the range of the u.v. doses used, the 40-fold increase represented an absolute increase of virulent phage mutants. Mutation to virulence in phage λ11 is a forward operator mutation whose location on the phage λ genetic map is known (Jacob and Wollman, 1954).

In order to interpret Jacob's data, I have carried out a few experiments with the working hypothesis that mutation to virulence in phage λ11 was due to the fact that the phage DNA had to replicate in a cell full of bacterial DNA breakdown products which could "poison" the incoming phage DNA development. Indeed, phage λ11 did not mutate to virulence in bacterial hosts carrying mutations in the *uvrA*, *uvrB* and *uvrC* genes. It is well known that such deficient bacteria do not break down their DNA after u.v. irradiation and are unable to excise pyrimidine dimers. So, lack of excision repair seemed to prevent u.v. host-induced phage mutagenesis (Devoret, 1965).

To assess this point, I wanted to construct complete isogenic strains differing only in the *uvrB* gene which governs the excision-repair function. Bacteria *gal uvrB* were transduced to either gal^+ $uvrB^+$ or gal^+ *uvrB* by phage P1 grown on C600. Surprisingly enough, phage λ11 could give rise to virulent mutants when grown in both u.v.-irradiated transduced strains. Mutation induction occurred at doses about five times lower in the *uvrB* bacteria than in the $uvrB^+$ ones, so that, in each case, mutation frequency to virulence appeared to be correlated with the fraction of non-excised pyrimidine dimers remaining in the host DNA.

In conclusion, host-induced u.v. mutagenesis of phage λ11 is dependent upon two factors: pyrimidine dimer excision repair and host-chromosome condition at the λ attachment site.

Weigle (1966) long ago postulated that u.v.-induced mutations in phage λ might arise from a recombinational event between phage λ and the host chromosome. My data do not prove the latter hypothesis very conclusively, but the fact is that the presence of non-excised pyrimidine dimers in a particular piece of bacterial chromosome in the region of the λ attachment site enhances the ability of phage λ11 to mutate to virulence.

Auerbach: How do you know that you were dealing with pyrimidine dimers? Could you photoreactivate the phage mutations?

Devoret: This has not yet been checked. The starting point of my experiments was to prove the idea that when an intact phage goes into a u.v.-irradiated cell whose DNA is broken down, it multiplies in an environment full of bacterial DNA pieces and may either incorporate them or mate with them. In five strains deficient in excision repair derived from the wild-type strain by nitrosoguanidine treatment, mutation induction did not occur. As soon as a particular piece of chromosome was introduced in three of the strains used, mutation induction ensued. Unexcised pyrimidine dimers certainly play a role but the host chromosome is also involved.

Auerbach: It must be something like that because irradiation of the host cell would not directly produce thymine dimers in phage; it could only produce them in the host chromosome. In the old experiments of the Texas school (Stone, Wyss and Haas, 1947) bacterial mutations were produced by treating the medium with u.v. These mutations cannot have been due to pyrimidine dimers.

Kubitschek: Evelyn Witkin (1968) has also proposed a recombinational process for mutation.

REFERENCES

Devoret, R. (1965). *C.r. hebd. Séanc. Acad. Sci., Paris,* **260,** 1510–1513.
Jacob, F. (1954). *C.r. hebd. Séanc. Acad. Sci., Paris,* **238,** 732–734.
Jacob, F., and Wollman, E. L. (1954). *Annls Inst. Pasteur,* **87,** 653–673.
Stone, W., Wyss, O., and Haas, F. (1947). *Proc. natn. Acad. Sci. U.S.A.,* **33,** 59.
Weigle, J. (1966). *Cold Spring Harb. Symp. quant. Biol.,* **31,** 226–235.
Witkin, E. M. (1968). *XII Int. Congr. Genet.,* Tokyo, **3.**

THE NATURE AND INFLUENCE OF ULTRAVIOLET AND HYDROXYLAMINE LESIONS IN NUCLEIC ACIDS AND THE ENZYMIC REPAIR OF THE FORMER

LAWRENCE GROSSMAN AND DANIEL M. BROWN

Graduate Department of Biochemistry, Brandeis University, Waltham, Massachusetts, and University Chemical Laboratory, Cambridge

THE effectiveness with which any mutagen can express itself in biological systems depends on the uniqueness and specificity of its chemical reactivity. Moreover its availability and survival within the cell, as well as the availability and specificity of its target, will determine its efficacy. It is not possible at present to assess the importance of these factors since they must vary from organism to organism. A schematic representation of the possible points of interaction of a mutagen with the informational pathway is given in Fig. 1.

The efficacy of a mutagen is obviously dependent on its gaining entrance into the cell. There are a number of instances in which a cell is resistant to chemical agents or base analogues by virtue of the selective inability of the cell to allow entry of such reagents. Moreover, even upon entry, mutagens such as hydroxylamine may be subject to degradative oxidation (Aleem, Lees and Lyrin, 1964) or reduction (Siegel, Crick and Moritz, 1964) by electron transport systems. Furthermore, the general viability of a cell is obviously dependent on the relative resistance of normal metabolic processes to such agents. For example, hydroxylamine is a potent inhibitor of the enzyme transaminase whose activity is vital in the general metabolic machinery of the cell.

The effects of either pyrimidine-specific reagents or base analogues are obscure, due to the paucity of relevant data. Thus the influence of 5-bromouracil-containing analogues of nucleotide coenzymes on polysaccharide and cell wall synthesis and other functions would have to be considered in the overall picture.

The most probable loci at which these agents assert their initial genotypic and phenotypic influences are either the template or the substrate levels in the recognition step implicit in replication, repair, transcription and

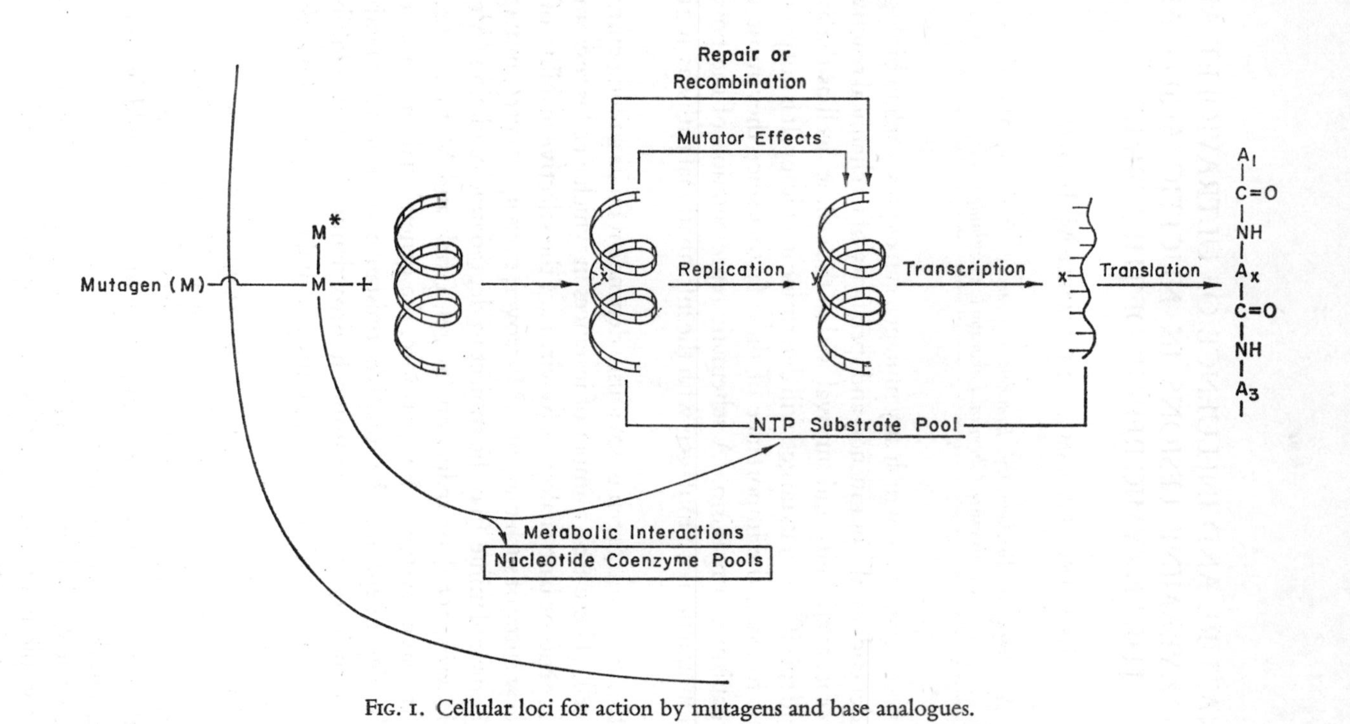

FIG. 1. Cellular loci for action by mutagens and base analogues.

translation. The underlying principles of recognition are relatively easy to examine and to interpret at all levels, with the exception of the last, namely the translational step. The protein-synthesizing system lends itself least readily to analysis not only because of its complexity but also, fundamentally, because of the large degree of code degeneracy and the three-to-one relationship between polynucleotide and polypeptide sequences. Since nascent polynucleotide structures are more amenable to sequence and structure analysis, polymerase systems have been used for the studies described here.

Investigations will be described which deal with the direct changes in recognition consequent on the modification of nucleic acid constituents. Most of the conclusions are based on experiments with the RNA polymerase from *Micrococcus luteus* (formerly *M. lysodeikticus*). Synthetic polydeoxyribonucleotides and ribonucleotides of known sequence and structure were used as templates for the RNA polymerase in which the introduction of specific lesions could be predictably controlled, quantitated and readily assayed for recognition changes (Fig. 2). In a second series of experiments,

Series I. Template modifications

Template	*RNA polymerase + required substrates*	Product
(*a*) Poly rC	GTP →	Poly rG
(*b*) Poly rU	ATP →	Poly rA
(*c*) Poly dAT	ATP → UTP	Poly rUA
(*d*) Poly dTG	ATP → CTP	Poly rCA
(*e*) Poly dAC	GTP → UTP	Poly rUG
(*f*) Poly rC(C)*	GTP → YTP†	Poly rGY

* Polynucleotide phosphorylase catalysed synthesis of random copolymer from CDP and CDP analogue.

† Purine ribonucleoside triphosphate which is complementary to C analogue.

FIG. 2. Series I type assay for modifications of template component pyrimidines.

similarly modified ribonucleoside triphosphates were prepared and substituted for normal pyrimidine ribonucleoside triphosphate substrates required for polymerization by this enzyme under the direction of unmodified polydeoxyribonucleotides of known sequence (Fig. 3). A third type of experiment, to assess modes of recognition, makes use of the nascent polyribonucleotides produced in the second series of experiments. In this

Series II. Nucleoside triphosphate substitutions

Template	RNA polymerase + required substrate	Product
(*a*) Poly dAT	ATP, (UTP)*	Poly r[U]A
(*b*) Poly dTG	ATP, (CTP)*	Poly r[C]A
(*c*) Poly dAC	GTP, (UTP)*	Poly r[U]G

* Replacement by:

(1) 5,6-Dihydrocytidine 5′-triphosphate

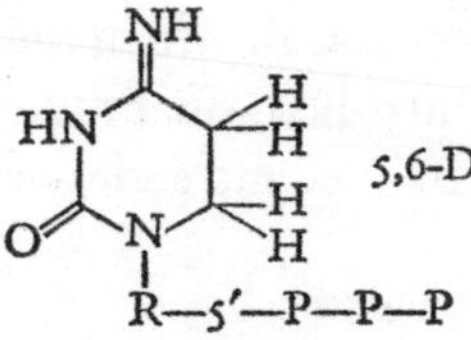

NH; HN; H; H; H; H; O; N; R—5′—P—P—P

(2) N^4-Hydroxycytidine 5′-triphosphate

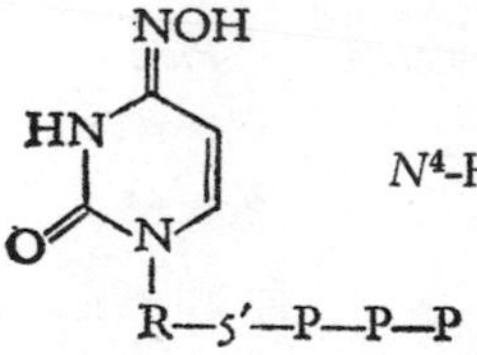

NOH; HN; O; N; R—5′—P—P—P

FIG. 3. Series II type assay for the utilization of modified RNA polymerase substrates.

case analogue polyribonucleotides of known sequence are employed as templates for the subsequent synthesis in the presence of normal nucleoside triphosphates (Fig. 4).

In all cases the nascent polyribonucleotides are examined for changes in sequence and primary structure involving the introduction of either modified pyrimidine nucleotides or replacement purine nucleotides which had been incorporated under the direction of templates containing lesions or those synthesized in the presence of substrate analogues.

MODIFICATION OF POLYCYTIDYLIC ACID

Fig. 5 provides a summary of much evidence obtained with templates, in this case polyribocytidylic acid, modified by a number of agents which

Series III. Second generation templating

1° *Template*	*RNA polymerase + required substrates*	2° *Template*	*RNA polymerase + required substrates*	*Product*
(*a*) Poly dAT	ATP → (DHCTP or N^4OHCTP)	Poly r [DHC or N^4OHC] A	UTP → (ATP or GTP)	Poly r [A or G] U
(*b*) Poly dTG	ATP → (DHCTP or N^4OHCTP) –	Poly r [DHC or N^4OHC] A	UTP → (ATP or GTP)	Poly r [A or G] U
(*c*) Poly dAC	GTP → (DHCTP or N^4OHCTP)	Poly r [DHC or N^4OHC] G	CTP → (ATP or GTP)	Poly r [A or G] C

FIG. 4. Series III type assay for the recognition properties of pyrimidine analogue components in nascent or second-generation templates for the RNA polymerase. DHCTP: dihydroxy-CTP; N^4OHCTP: N^4-hydroxy-CTP.

effectively reduce a proportion of its constituent cytosine residues at the 5,6-double bond.

The RNA polymerase of *M. luteus* is able to utilize either DNA or RNA as a template for RNA synthesis. Moreover, certain polyribonucleotides may substitute for the latter when the appropriate ribonucleoside triphosphate substrates are provided. In all cases, these templates direct synthesis with a great deal of fidelity, the fundamental basis of which appears to be Watson-Crick hydrogen-bonding complementarity. In the specific cases illustrated (Fig. 5), polycytidylic acid (poly rC) acts as a specific template for the polymerization of GTP to polyguanylic acid (poly rG). Modification of such a template by a variety of agents which effectively reduce the

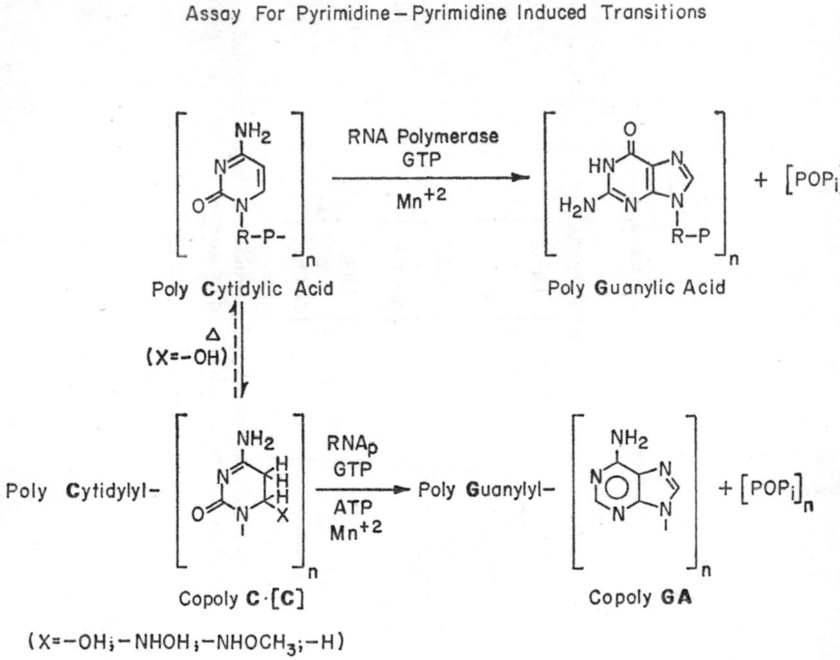

Assay For Pyrimidine—Pyrimidine Induced Transitions

RNA Polymerase, GTP, Mn^{+2}

Poly Cytidylic Acid → Poly Guanylic Acid + $[POP_i]_n$

Δ (X=–OH)

RNAp, GTP, ATP, Mn^{+2}

Poly Cytidylyl– (Copoly C·[C]) → Poly Guanylyl– (Copoly GA) + $[POP_i]_n$

(X=–OH; –NHOH; –$NHOCH_3$; –H)

FIG. 5. Summary of RNA polymerase (RNAp) reactions directed by a modified polycytidylic acid template.

cytosine 5,6-double bond results in the formation of a copolymer which specifically directs the synthesis of polyguanyladenylate (copoly rGA). The ATP incorporated internally into the nascent copoly rGA is specifically directed by the modified cytosine residues. The various adducts which are formed all have in common the reduction of the 5,6-double bond. The elements of water added at this particular site as a consequence of ultraviolet irradiation are readily removed by heat, thereby restoring the original cytosine structure. Reversal of the ATP incorporation directed by irradiated poly C is accomplished under conditions of dehydration, thereby implicating this particular unstable photoproduct in a change in recognition (Ono, Wilson and Grossman, 1965).

The photoproducts expected in DNA (*E. coli*) irradiated with 260 nm ultraviolet light and the relative proportion of each in a given population of photoproducts are given in Fig. 6. The influence of these specific photoproducts on the mutational and lethal events accompanying the irradiation will become more apparent later.

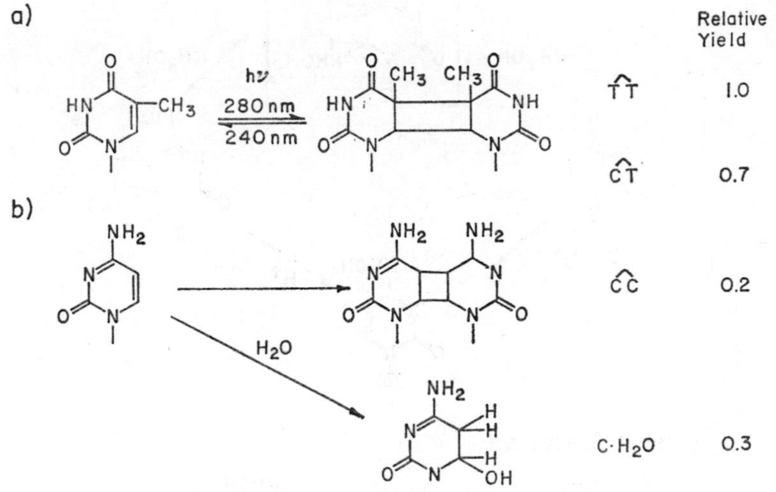

	Relative Yield
T̂T	1.0
ĈT	0.7
ĈC	0.2
C·H_2O	0.3

FIG. 6. Nature and proportion of photoproducts found in ultraviolet-irradiated DNA.

The relevance of cytosine hydration to the mutational events associated with ultraviolet irradiation will, by a process of elimination, become more apparent from the ensuing discussion of the properties of reduced thymine and uracil residues. Moreover, the direct quantitative estimation of the unstable cytosine photohydrate in DNA provides insight into its significant place in any population of photoproducts in irradiated DNA (Grossman and Rodgers, 1968).

ALTERATION BY HYDROXYLAMINE

A recent review by Phillips and Brown (1967) discusses the chemistry of hydroxylamine. The adducts of cytosine with hydroxylamine and with O-methylhydroxylamine, which share structural similarities with the photohydrate of cytosine, cause similar mutagenic changes in phage systems and corresponding transitional changes *in vitro* (Phillips and co-workers, 1965) (see Fig. 7). Experiments with [^{14}C]-*O*-methylhydroxylamine permitted indirect measurement of the proposed monoadduct in Fig. 7 as well as allowing for measurements of the number of such adducts formed in poly C. These data, in conjunction with a direct assessment of the number of ATP molecules incorporated after treatment with this agent,

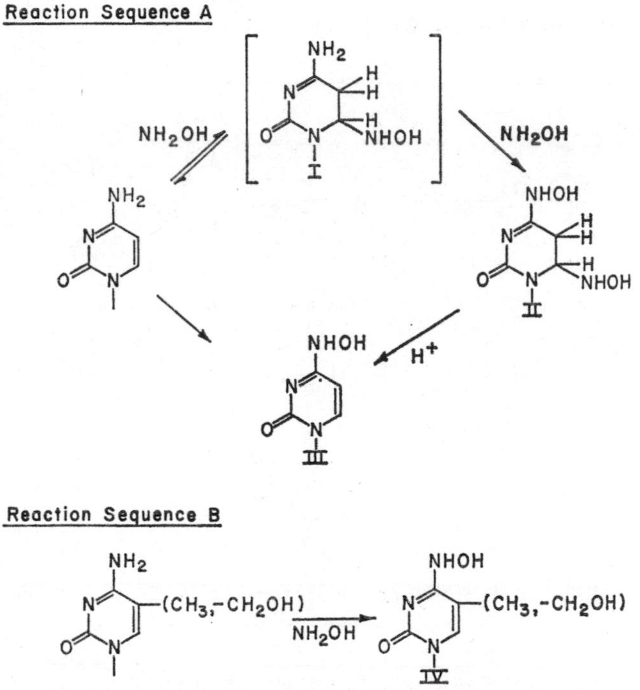

FIG. 7. Summary of hydroxylamine reaction products of cytosine and 5′-substituted cytosines.

allowed an estimate to be made of the efficiency with which the RNA polymerase copied the adduct of cytosine as if it were uracil or thymine. The high efficiency which was observed in such experiments (1 mole of ATP incorporated/mole of monoadduct formed in the template) is in accord with the genetic experiments of Brenner, Stretton and Kaplan (1965) and Brenner and co-workers (1967), and lends some credence to the significance of such *in vitro* experiments in mutagenesis (Phillips, Brown and Grossman, 1966).

The generalization that reduction of the 5,6-double bond is exclusively responsible for transition mutations cannot fully account for the mutagenic influence of hydroxylamine and of u.v. light. Thus the reaction of hydroxylamine with 5-substituted cytosines (reaction sequence B in Fig. 7) represents, according to Janion and Shugar (1965*a*, *b*), the primary hydroxylamine reaction with these particular pyrimidines. Hydroxylamine is mutagenic for bacteriophage T4 and presumably the N^6-exchange product is the primary product of such a reaction.

Copolymers of cytidylic and N^4-hydroxycytidylic acids have been prepared and also act as effective templates for the synthesis of copoly (AG) as catalysed by the RNA polymerase (Banks, Brown and Grossman, 1969).

The relatively high degree of efficiency with which RNA polymerase transcribes these adducts and the unmistakable specificity of the C to U(T) transition induced by them might be explained by invoking one of two general chemical mechanisms. We assume that in order to explain this change in recognition by the RNA polymerase it is necessary that the adducts in question have hydrogen-bonding properties potentially similar to those of uracil or thymine. Thus N^3 should be protonated either by virtue of a change in the tautomeric form or by addition of a proton (to give a cation). The pK_a of cytosine is 4·2 and reduction leads to the stronger dihydro base having a pK_a of 6·2. It would be expected that the probability of such a reduced species being protonated at the pH of the enzyme reactions would be increased from 1 in 2000 to 1 in 20. The most striking change arising from reduction of the 5,6-double bond is in the tautomeric constant (K_t). The K_t of 10^5 for cytosine, in the direction of the amino form, is shifted down to 25 for 5,6-dihydrocytosine in water and to less than 1 in chloroform (Brown and Hewlins, 1968; Brown, Hewlins and Schell, 1968). In order to test the contributions of pK_a and K_t changes an approach somewhat different to that already described was pursued.

SUBSTRATE MODIFICATIONS

When he applies experimental results obtained from modified templates the investigator is subject to certain difficulties inherent in quantitating the actual number of modifications, since the polymer used for analysis is usually not the same as that employed as a template. Moreover, a determination of the frequency with which a modified base behaves as its homologue (in this case as cytosine) rather than a transitional pyrimidine is difficult to estimate directly.

The experimental approach used to determine some aspects of the mechanism governing such recognition changes is outlined in Fig. 3. In such experiments enzymically synthesized polymers of defined sequence are employed as templates for this enzyme reaction, whose requirement for either UTP or CTP is absolute. Substitution of the required pyrimidine nucleoside triphosphate by a purified analogue provides a definitive and quantitative estimate of the ability of the analogue to replace one of the required pyrimidine nucleoside triphosphates. Moreover, nearest-

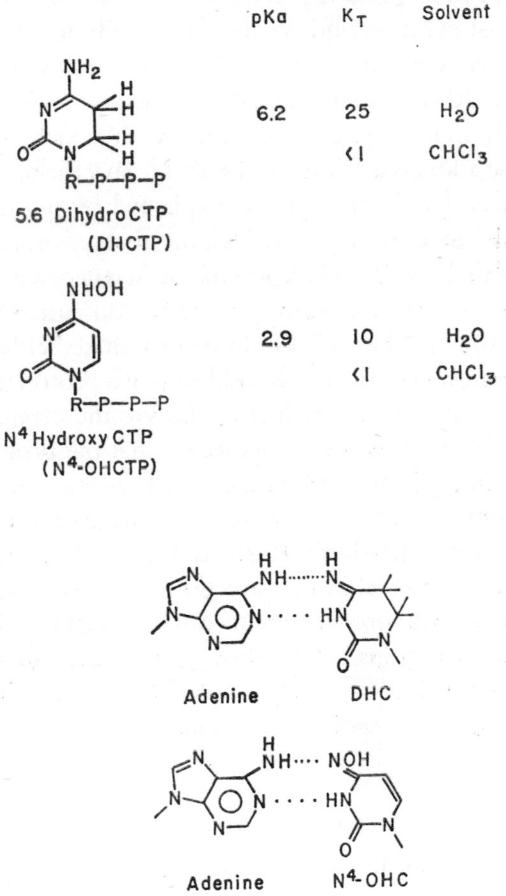

	pKa	K_T	Solvent
5.6 DihydroCTP (DHCTP)	6.2	25	H_2O
		<1	$CHCl_3$
N^4 Hydroxy CTP (N^4-OHCTP)	2.9	10	H_2O
		<1	$CHCl_3$

Adenine DHC

Adenine N^4-OHC

FIG. 8. Structure and properties of analogues replacing UTP or CTP.

neighbour frequency analyses of the nascent polyribonucleotide provide unequivocal proof of the sequence of such a polymer.

Substrate analogues of essentially different structures were chosen with very similar tautomeric constants but possessing very different pK_a values (Fig. 8). These analogue substrates were used to satisfy the CTP and UTP requirement in the reaction sequences already depicted. The data described in Table I were obtained from both kinetic experiments and substrate saturation data. Routine nearest-neighbour frequency analyses were performed on the nascent polyribonucleotides in which α-^{32}P-labelled analogue nucleoside triphosphates had been incorporated. In all cases examined, the sequence of the newly synthesized polyribonucleotide was essentially complementary to that of the polydeoxyribonucleotide employed as a template.

The results in Table I provide five fairly illuminating aspects of the utilization of these analogue nucleoside triphosphates. (1) It appears from the data that both dihydro-CTP and N^4-hydroxy-CTP can effectively replace both CTP and UTP. From early experiments (Grossman, Kato and Orce, 1966) we knew that dihydro-CTP was not able to replace UTP or CTP when a polyribonucleotide was used as a template, whereas DNA and poly dAT were effective templates for utilization of this analogue. This unexplained specificity for the "natural" templates, as will be seen later, had more to do with the sequence of the purine nucleotides in the polyribonucleotides than with the nature of the carbohydrate residue. (2) These data provide clear support for the transitions observed when similar modifications were introduced into the templates by either u.v. or hydroxylamine. Again, transitions in recognition are observed, but now at the substrate level. The biological implication of these data is that modification by mutagenic agents may occur at the polymer as well as the substrate level. That both these substrates are preferentially utilized as CTP lends credence to the notion that (3) the alteration in the tautomeric constant is the most likely cause for change in recognition. The analogues chosen for these experiments generally had a different structure and pK_a, but were as nearly alike in tautomeric constant as possible. In these circumstances, we are forced to conclude that their identical efficiencies in incorporation, regardless of template, imply that the tautomeric state is the dominant characteristic which contributes to changes in recognition.

However, the utilization of these modified substrates as either CTP or UTP is not strictly limited by the inherent hydrogen-bonding properties of the analogues as judged only from consideration of the K_t and pK. Surprisingly the sequence of the template itself appears to play an important directing role. Thus if the adenine residues alternate with either thymine or cytosine in the template this appears to make a difference to whether the analogue can replace UTP. Moreover, homopolymers (last column of Table I) apparently cannot direct the utilization of these analogues at all. These unusual findings clarify earlier data which show that polyribonucleotides containing adenine exclusively will not direct the incorporation of

TABLE I

THE EFFICIENCY (%) OF ANALOGUE NUCLEOSIDE TRIPHOSPHATE INCORPORATION WITH TEMPLATES OF DEFINED SEQUENCE

	Template reaction			
	Poly dpApTpApTp	Poly dpApCpApCp	Poly dpGpTpGpTp	Poly dpApApApAp
	ATP ↓ (UTP)	GTP ↓ (UTP)	ATP ↓ (CTP)	↓ (UTP)
Analogue	Poly rpApUpApUp	Poly rpGpUpGpUp	Poly rpCpApCpAp	Poly rpUpUpUpUp
5,6 Dihydro-CTP	12·8	30·4	66·6	0
N^4-OH-CTP	12·1	25·7	65·4	0
UTP	100	100	0	100
CTP	0	0	100	0

dihydro-CTP. Thus, the mechanisms of these recognition changes cannot be accurately defined solely in terms of the hydrogen-bonding properties of the substrate. The influence of the sequence of the templates on the effect of hydrogen-bonding properties of the nitrogenous bases directing the incorporation of these analogues is clearly not understood at this time.

The final portion of the data presented in Table I concerns the relative effectiveness of these analogues when utilized either as CTP or UTP. It is clear in all cases that there is a greater likelihood that the analogue will be utilized as CTP rather than UTP. The tautomeric constants for the N^1 methyl derivatives indicate that the amino form is more dominant in aqueous media, and the imino derivative predominates in solvents such as chloroform. Since the environment in and around the pyrimidine moieties in templates and substrates is not defined, it may tentatively be assumed from these data that the environment corresponds to a less hydrophobic state than that of such non-aqueous solvents as chloroform.

INDUCED REITERATION

Polyuridylic acid (poly U) has been examined analogously to poly C. Poly U specifically directs the polymerization of ATP forming polyadenylic acid and inorganic pyrophosphate. The techniques for modification of the uridylic acid residues in poly U are analogous to those used in the formation of polycytidylates. Fig. 9 depicts the overall pattern obtained with modifications of such uracil residues (Adman and Grossman, 1967). The reduction of the 5,6-double bond of the uridylate moieties in poly U results in the formation of a copolymer of uridylate with a reduced uracil residue as an "unnatural" block. The block to transcription leads to a completely different expression of action from that seen with the poly C series. In this case, reduction of the uracil residues leads to a loss of recognition and concomitant interference with the action of the RNA polymerase. The blocking of this enzyme by these reduced uracil residues is not a passive effect, but rather leads to reiteration of poly A synthesis. Reiteration of poly A synthesis is also observed with various copolymer templates containing cytosine, adenine or guanine, in addition to uracil. In the absence of the complementary nucleoside triphosphates, reiteration is observed. For example, copolymers containing uracil and guanine normally direct the synthesis of copoly AC from ATP and CTP. If the latter is withheld from the reaction mixture, poly A reiteration occurs. An expression of such reiteration is an unabated increase in the rate of synthesis of poly A when compared to the normal. Synthesis of poly A only levels

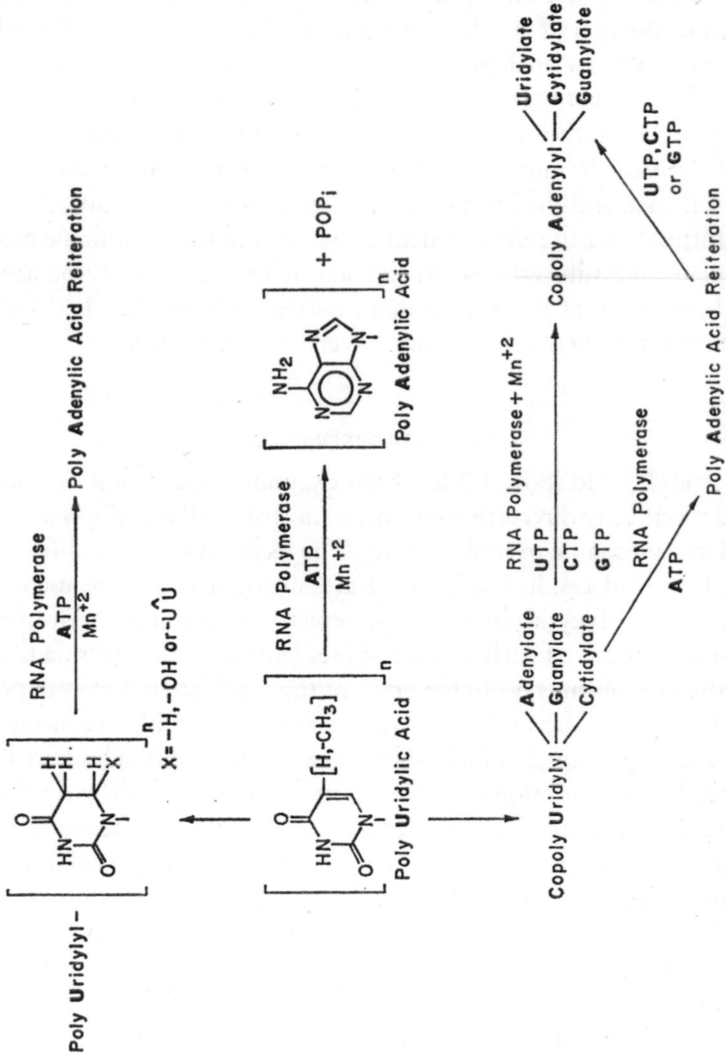

FIG. 9. Summary of effects of uracil modifications and replacement on poly U template activity.

off when the substrate ATP is exhausted. The difference seen between a reiterative reaction and a normal transcriptive reaction is that the reiteration reaction is substrate-limited whereas transcription is normally template-limited. In addition to these hyperkinetic phenomena, the average chain length of the nascent poly A produced under reiteration conditions is considerably greater than the chain length of the original template. In normal transcription the average mean chain length of the nascent polynucleotide invariably equals that of the original template.

The reiteration observed with natural copolymers when complementary nucleoside triphosphates are withheld can be reversed by very low levels of the missing nucleoside triphosphate. Subsequent addition of such nucleoside triphosphates will not relieve induced reiteration if some of the uracil

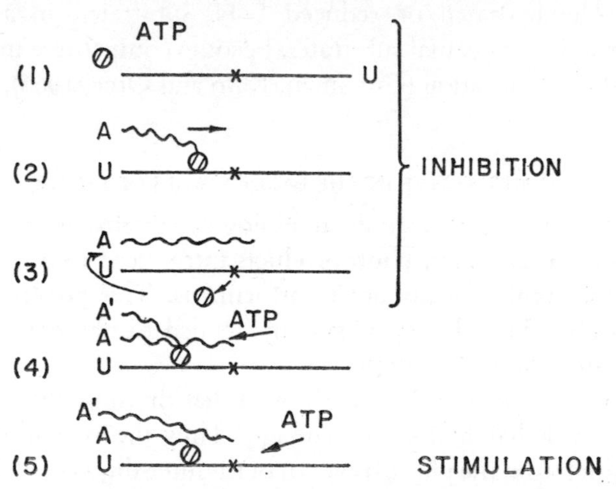

FIG. 10 Hypothetical mechanism proposed for reiteration by the RNA polymerase.

residues are either reduced, photohydrated, or photodimerized. It would appear, then, that reduction of uracil or thymine residues leads to a loss of recognition, possibly through lack of hydrogen-bonding ability. These data are in agreement with the findings of Cerutti, Miles and Frazier (1966), who examined the hydrogen-bonding properties of partially or fully reduced polyuridylate with poly A. Moreover, reduction of the uracil residues in triplets results in loss of binding by such messengers to ribosomes in the presence of tRNA. Fig. 10 summarizes what we suggest may be the general mechanism of induced reiteration: an enzyme molecule randomly initiates transcription on a polyuridylic acid template, it proceeds to synthesize poly A reading in the 3′ to 5′ direction, and the nascent poly A is hydrogen-bonded to the template up to the point of the lesion. The enzyme leaves the template at this point, binds to the double-stranded portion of the partially replicated template, continues to read up to the point of the block, and rather than stop proceeds to re-read the template reiteratively. Statistical analyses of a larger number of copolymers reveal that the enzyme re-reads three or four nucleotides before this block in a completely reiterative fashion.

REDUCTION OF UTP

Reduction and u.v. hydration or dimerization of UTP substrates for the polymerase reaction also lead to a total loss of recognition. These findings parallel the observations on similar modifications of uracil residues in templates. The hydrated or reduced UTP substrates, in addition to becoming inactive as normal substrates, become competitive inhibitors of the normal UTP utilization (Grossman, Kato and Orce, 1966).

ENZYMES ACTIVE IN THE EARLY STAGES OF REPAIR

Attempts to extend earlier studies to biological systems meant correlating the number of specific u.v. photoproducts introduced into transforming DNA with the number of mutant transformants. This proved to be virtually impossible since the repair systems tended to obliterate or correct the lesions which were introduced.

Setlow and Setlow (1962, 1963) showed that thymine-thymine dimers in DNA cause lethal biological damage. Thus the overall dark-repair process (distinct from the photo-reversal of thymine dimers—Kelner, 1949) must be capable of dimer removal or reversal. Enzymic excision of these photoproducts was suggested by the work of Boyce and Howard-Flanders (1964) and Setlow and Carrier (1964), who simultaneously demonstrated that [^{3}H]thymine dimers were released from the DNA of u.v.-resistant bacteria during a post-irradiation recovery period.

Dark repair is not confined to u.v.-induced damage, but also includes lesions produced by such diverse agents as mitomycin, X-rays, nitrous acid and certain monofunctional and difunctional alkylating agents (Howard-Flanders and Boyce, 1966). Therefore, enzymes active in the initial stages of the repair system are not necessarily specific for primary chemical changes in the DNA, but may recognize the secondary effects, for example, localized regions of distortion caused by dimer formation. Strauss (1962), Elder and Beers (1965), Moriguchi and Suzuki (1966), and Carrier and Setlow (1966) have all reported the existence, in crude extracts of *M. luteus*, of enzymic activity which specifically degrades u.v.-irradiated double-stranded DNA.

We have been interested in isolating enzymes which can recognize localized distortions in DNA structure. *M. luteus* was chosen as the source because of its low levels of endogenous nuclease. This permits the detection of small amounts of distortion-specific endonucleases and exonucleases. The properties of two enzymes which together excise photoproduct-containing regions of u.v.-irradiated duplex DNA are described here. The first enzyme is an endonuclease that causes a single phosphodiester bond break to occur near each pyrimidine dimer, resulting in the formation of a single-stranded region. A similar enzyme has been reported by Nakayama and co-workers (1967) and Takagi and co-workers (1968). The second enzyme, an exonuclease, is capable of degrading exposed single-stranded regions resulting from incision. The combined action of these two enzymes results in the release of acid-soluble nucleotides.

PROPERTIES OF U.V. ENDONUCLEASE (INCISION ENZYME)

The hypothesis that the u.v. endonuclease makes a single-strand incision rather than double-strand cleavage is supported by the evidence that when DNA treated with the enzyme is subjected to ultracentrifugation there is a rapid and significant decrease in the sedimentation coefficient under alkaline conditions. The finding that a change in the sedimentation coefficient only occurred in alkaline solution is indicative of single-stranded scission.

The assay for the endonuclease is based on its ability to hydrolyse phosphodiester bonds in u.v.-irradiated double-stranded [^{32}P]DNA. By measuring the appearance of ^{32}P-labelled phosphomonoester groups, using bacterial alkaline phosphatase, it is possible to determine quantitatively the number of phosphodiester bonds broken. Incision by the endonuclease produces a single-stranded region which is subsequently recognized by the u.v. exonuclease. Actual nucleotide release requires the presence of both enzymes.

The endonuclease has been purified more than 5000-fold and is entirely free of contaminating exonuclease activity. The enzyme has a molecular weight of 15 000, and does not require Mg^{2+} but is activated by it. The pH optimum is between 6·5 and 7·5. Fig. 11 shows that the endonuclease requires u.v.-irradiated double-stranded DNA and that its activity is dose-dependent.

The high efficiency of the endonuclease for photoproduct-induced distortions is also indicated in Fig. 11. In the region between 10^4 and 10^5

erg/mm² at 280 nm, where thymine-thymine dimer levels have been determined independently by chromatographic methods, one phosphodiester bond is broken for each dimer. According to the work of Setlow and Carrier (1966), approximately 94 per cent of all pyrimidine-pyrimidine dimers formed at 280 nm contain at least one thymine residue. Therefore, the enzyme not only exhibits a specificity for u.v.-induced distorted regions, but also quantitatively produces a single incision in each of these regions. Preliminary results indicate that enzymic attack occurs on the strand containing the photoproducts rather than on the complementary strand.

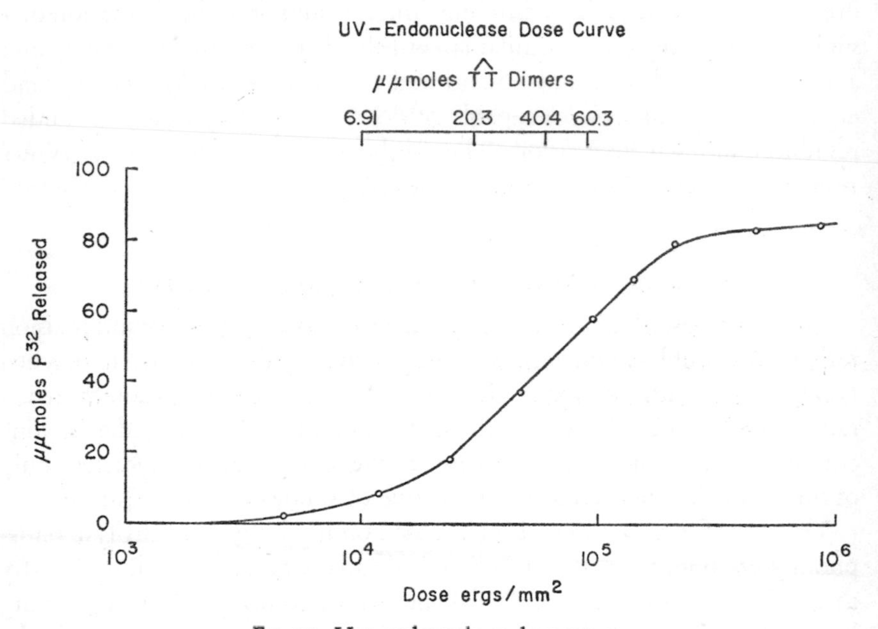

FIG. 11. U.v. endonuclease dose curve.

PROPERTIES OF U.V. EXONUCLEASE (EXCISION ENZYME)

The u.v. exonuclease has been purified more than 1000-fold and is also free of conflicting nucleases. The absolute dependence of enzymic activity on Mg^{2+} cannot be satisfied with other divalent cations (Ca^{2+}, Mn^{2+} or Zn^{2+}). The enzyme with an approximate molecular weight of 17 000 is stable between pH 6·5 and 8·0 and is specifically activated by ammonium sulphate (0·001–0·01 M).

The exonuclease is assayed by measuring the conversion of u.v.-irradiated denatured [³²P]DNA into acid-soluble nucleotides. During purification a comparison is made between both irradiated and unirradiated denatured DNA as substrate. A time course of the enzymic reaction demonstrates that the purified u.v. exonuclease does not discriminate between denatured [^{32}P]DNA and irradiated denatured [^{32}P]DNA. However, a second exonuclease is present in *M. luteus* which actively hydrolyses denatured DNA but is inhibited by irradiated denatured DNA.

In addition, the u.v. exonuclease is able to degrade 5′-*O*-phosphoryl-oligothymidylates of various chain lengths. In contrast to exonuclease I from *Escherichia coli* (Lehman, 1963), u.v. exonuclease can completely hydrolyse 5′-*O*-phosphorylthymidylyl-thymidine. These oligonucleotides have not been irradiated.

The number of photoproducts in denatured DNA produced at 280 nm does not affect u.v. exonuclease activity. In contrast, snake venom phosphodiesterase and *E. coli* exonuclease I, which are similar enzymes, are markedly inhibited by irradiated DNA. The fact that u.v. exonuclease activity is not depressed by a photoproduct containing DNA strongly suggests that this enzyme is involved in the dark repair system.

SIZE OF THE EXCISED REGIONS

The number of phosphodiester bonds broken, relative to the total number of nucleotides released by exonuclease action, gives some indication of the size of the excised region. Sequential use of endonuclease and exonuclease results in the release of five nucleotides for each incision event.

Some indication of the composition of the damaged region can be obtained by using irradiated DNA labelled with [^{3}H]thymidine as substrate. Under these circumstances three thymine-containing residues, on average, are released for every phosphodiester bond broken by the u.v. endonuclease. Since a total of five nucleotides are released, dimerization events probably occur more often in regions containing runs of thymine rather than in isolated dinucleotide sequences.

INVOLVEMENT OF NUCLEASES IN BIOLOGICAL REPAIR

Even though the combined action of these two nucleases is consistent with current models for the excision of damaged DNA *in vivo*, it must be established that these enzymes are in fact part of the cellular repair process. In order to clarify this point, a series of u.v.-sensitive (*uvr*$^{-}$) mutants of *M. luteus* were isolated.

Cells of *M. luteus* (ATCC No. 4698) were treated with *N*-methyl-*N*′-nitro-*N*-nitrosoguanidine, according to the method of Adelberg, Mandel and Chen (1965), or were plated on brain-heart infusion agar and exposed

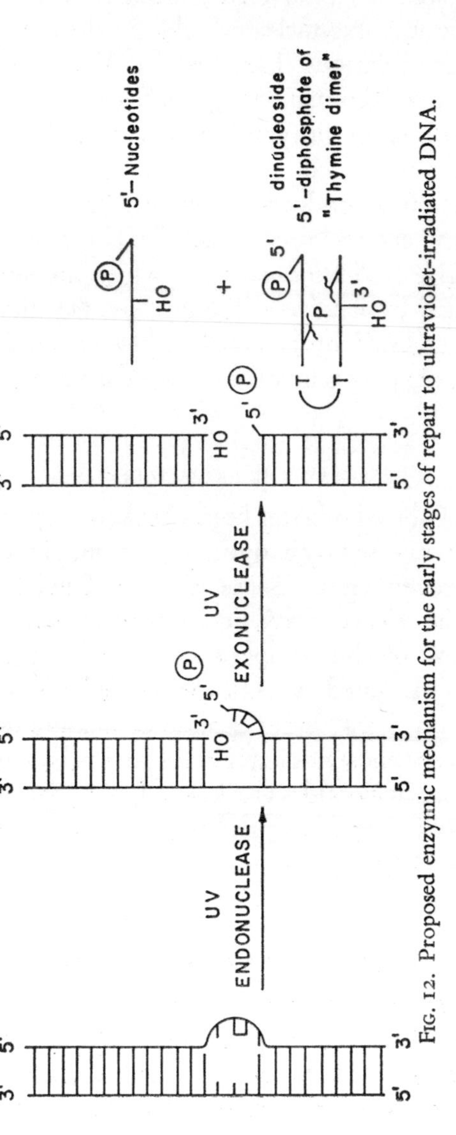

FIG. 12. Proposed enzymic mechanism for the early stages of repair to ultraviolet-irradiated DNA.

to 10^5 erg/mm^2 of u.v. Mutants sensitive to 0·05 mg of mitomycin C/ml were selected by replica-plating (Okubo and Romig, 1965) and examined for simultaneous sensitivity to u.v.- and to X-radiation. All u.v.-sensitive mutants were examined for repair of u.v.-irradiated bacteriophage B4 (*hcr*), sensitivity to 0·05M-methyl methanesulphonate (Searashi and Strauss, 1965), endonuclease activity, and exonuclease activity. The ability of cellular extracts to degrade u.v.-irradiated *Bacillus subtilis* transforming DNA provided a simple assay for the combined action of the two enzymes.

Extracts from one of the u.v.-sensitive *hcr*$^-$ mutants, ML-7, which was also sensitive to X-rays, were found to contain no detectable levels of u.v. endonuclease activity, in contrast to wild-type extracts. In a balance experiment cells from both these strains were labelled with [^{3}H]thymidine and u.v. irradiated; the levels of radioactivity were then determined in the DNA and the acid-soluble fractions while the cells were undergoing post-irradiation dark recovery. This type of experiment clearly indicates the differences between the *uv*$^+$ wild type, which is capable of excision, and the *uv*$^-$ mutant ML-7, which removes thymidine at a significantly lower rate, in keeping with its reduced endonuclease activity.

A possible model for dark repair is shown in Fig. 12. Our evidence suggests that the nucleases described could be responsible for the early events in dark repair. The incision step may not be required for the repair of DNA damaged by agents which themselves produce single-strand breaks which would negate the requirement for the incision (u.v. endonuclease) step.

SUMMARY

The modification of cytosine residues in polyribonucleotides by ultraviolet irradiation or hydroxylamine results in a C→U(T) type transition *in vitro* when measured by the RNA polymerase. A similar transition is observed when modified nucleoside triphosphates are employed as substrates with unmodified polynucleotide templates. In those cases studied the transition in recognition by these agents is attributable to a large depression in the tautomeric constants of cytosine from 10^5 to close to unity.

Similarly induced modifications to uracil and thymine result in a loss of recognition at the template level and also at the level of the substrates. This loss of recognition has been ascribed to the loss of hydrogen bonding capabilities of effectively reduced uracil and thymine ring structures.

Two enzymes have been purified from *M. luteus* which together excise photoproducts from u.v.-irradiated DNA. The first enzyme is an endonuclease that produces single-strand breaks in irradiated DNA but does not attack native or denatured DNA. Nucleotide release is dependent on the presence of a second enzyme, a u.v. exonuclease, which hydrolyses single-stranded regions of incised irradiated DNA. A singular property of this exonuclease is that its activity is not inhibited by the presence of photo-products in polynucleotide chains. The combined action of these two enzymes results in the release of approximately four to five nucleotides for each phosphodiester bond broken. The majority of the released nucleotides contain thymine.

Acknowledgements

The authors wish to acknowledge the contributions of G. R. Banks, D. Streeter, S. Kushner, J. Kaplan and I. Mahler in the experiments which have yet to be published. In addition, we wish to recognize the agencies which provided the much-needed financial support: National Institutes of Health (GM 15881), National Institutes of Health Career Development Award (GM 4845), National Science Foundation (GB 6208), Atomic Energy Commission (AT (30-1) 3449), and the American Cancer Society (E-490). This is contribution No. 644 of the Graduate Department of Biochemistry, Brandeis University.

REFERENCES

ADELBERG, E. A., MANDEL, M., and CHEN, G. C. C. (1965). *Biochem. biophys. Res. Commun.*, **18**, 788.

ADMAN, R., and GROSSMAN, L. (1967). *J. molec. Biol.*, **23**, 417.

ALEEM, M. I. H., LEES, H., and LYRIN, R. (1964). *Can. J. Biochem. Physiol.*, **42**, 489.

BANKS, G. R., BROWN, D. M., and GROSSMAN, L. (1969). In preparation.

BOYCE, R. P., and HOWARD-FLANDERS, P. (1964). *Proc. natn. Acad. Sci. U.S.A.*, **51**, 293.

BRENNER, S., BARNETT, L., KATZ, E. R., and CRICK, F. H. C. (1967). *Nature, Lond.*, **213**, 449.

BRENNER, S., STRETTON, A. O. W., and KAPLAN, S. (1965). *Nature, Lond.*, **206**, 695.

BROWN, D. M., and HEWLINS, M. J. E. (1968). *J. chem. Soc. C.*, 2050.

BROWN, D. M., HEWLINS, M. J. E., and SCHELL, P. (1968). *J. chem. Soc. C.*, 1925.

CARRIER, W. L., and SETLOW, R. B. (1966). *Biochim. biophys. Acta*, **129**, 318.

CERUTTI, P., MILES, H. T., and FRAZIER, J. (1966). *Biochem. biophys. Res. Commun.*, **22**, 466.

ELDER, R. L., and BEERS, R. F. JR. (1965). *J. Bact.*, **90**, 681.

GROSSMAN, L., KATO, K., and ORCE, U. L. (1966). *Fedn Proc. Fedn Am. Socs exp. Biol.*, **25**, 276.

GROSSMAN, L., and RODGERS, E. (1968). *Biochem. biophys. Res. Commun.*, **33**, 975.

HOWARD-FLANDERS, P. and BOYCE, R. P. (1966). *Radiat. Res.*, Suppl. 6, 156–184.

JANION, C., and SHUGAR, D. (1965*a*). *Acta biochim. pol.*, **12**, 338.

JANION, C., and SHUGAR, D. (1965*b*), *Biochem. biophys. Res. Commun.*, **18**, 617.

KELNER, A. (1949). *Proc. natn Acad. Sci. U.S.A.*, **35**, 73.

LEHMAN, I. R. (1963). *Prog. Nucleic Acid Res.*, **2**, 91.

MORIGUCHI, E., and SUZUKI, K. (1966). *Biochem. biophys. Res. Commun.*, **24**, 195.

NAKAYAMA, H., OKUBO, S., SEKIGUCHI, M., and TAKAGI, Y. (1967). *Biochem. biophys. Res. Commun.*, **27**, 217.

OKUBO, S., and ROMIG, W. R. (1965). *J. molec. Biol.*, **14**, 130.

ONO, J., WILSON, R .G., and GROSSMAN, L. (1965). *J. molec. Biol.*, **11**, 600.

PHILLIPS, J. H., and BROWN, D. M. (1967). *Prog. Nucleic Acid Res.*, **7**, 349.

PHILLIPS, J. H., BROWN, D. M., ADMAN, R., and GROSSMAN, L. (1965). *J. molec. Biol.*, **12**, 816.

PHILLIPS, J. H., BROWN, D. M., and GROSSMAN, L. (1966). *J. molec. Biol.*, **21**, 405.
SEARASHI, T., and STRAUSS, B. (1965). *Biochem. biophys. Res. Commun.*, **20**, 680.
SETLOW, J. K., and SETLOW, R. B. (1963). *Nature, Lond.*, **197**, 560.
SETLOW, R. B., and CARRIER, W. L. (1964). *Proc. natn Acad. Sci. U.S.A.*, **51**, 226.
SETLOW, R. B., and CARRIER, W. L. (1966). *J. molec. Biol.*, **17**, 237.
SETLOW, R. B., and SETLOW, J. K. (1962). *Proc. natn Acad. Sci. U.S.A.*, **48**, 1250.
SIEGEL, L. M., CRICK, E. M., and MORITZ, K. J. (1964). *Biochem. biophys. Res. Commun.* **17**, 125.
STRAUSS, B. S. (1962). *Proc. natn Acad. Sci. U.S.A.*, **48**, 1670.
TAKAGI, Y., SEKIGUCHI, M., OKUBO, S., NAKAYAMA, H., SHIMADA, K., YASUDA, S., MISHIMOTO, T., and YOSHIHARA, H. (1968). *Cold Spring Harb. Symp. quant. Biol.*, **33**, 219.

DISCUSSION

Bridges: How unstable are cytosine hydrates *in vivo*?

Grossman: The stability of cytosine hydrates *in vitro* depends on temperature, pH and the conformation of the DNA itself. The rate constants for the dehydration of cytosine hydrates is between 10^{-3} to 10^{-2} min^{-1} at pH 7 (Johns, LeBlanc and Freeman, 1965). The quantum yield for the formation of cytosine hydrates varies according to the conformation of the DNA: for example, it is $1 \cdot 6 \times 10^{-3}$ moles per microeinstein in native DNA and considerably higher ($5 \cdot 1 \times 10^{-3}$ moles per microeinstein) for denatured DNA (Grossman and Rodgers, 1968). It is virtually impossible to obtain such numbers for the formation of this photoproduct from experimentation *in vivo*.

Bridges: What sort of half-life would you expect at 45°?

Grossman: In polyribocytidylic acid (poly C) the half-life for cytosine hydrates at 45° would be no more than about 10 minutes (Ono, Wilson and Grossman, 1965).

Bridges: So one could test for the involvement of such lesions by looking at the effect of temperature during a non-growing period after u.v.?

Grossman: This might be feasible. However, since there are so many different u.v. photoproducts whose formation and survival are dependent on temperature, pH and DNA structure, the alteration of one species must influence, in turn, the presence of the others. It would be much less equivocal to employ agents which cause only single chemical changes with one type of nitrogenous base. For example, hydroxylamine or u.v. irradiation in the presence of sensitizers would allow for the biological effects in a more direct manner.

Rörsch: Mr. Hout in our laboratory has also purified the endonuclease and exonuclease involved in the excision of dimers, though he has not achieved such a high purification as you mentioned. I am surprised to learn that your mutant that lacks the u.v. specific endonuclease is u.v.-sensitive and lacks host cell reactivation. My collaborator Mr. C. A. van Sluis has also isolated two mutants of *M. lysodeikticus* (strain MBL 9 and strain MBL 10) lacking the endonuclease activity but *hcr*$^+$ and showing the same u.v. sensitivity as the wild type. A similar mutant was described by Okubo and co-workers (1967).

Magni: In what respect are they mutant?

Rörsch: They lack the endonuclease activity, are unable to excise pyrimidine dimers from u.v.-irradiated DNA *in vitro* and are unable to restore the biological activity of u.v.-irradiated phage DNA *in vitro* (Rörsch, van Sluis and van de Putte, 1966).

Grossman: Those observations are not unique. Dr. Inga Mahler in our laboratory has isolated several classes of transformants of *M. luteus* which lack u.v. endonuclease and are resistant to ultraviolet (*uv*$^+$). These data clearly implicate other mechanisms of repair which do not involve the excision of thymine dimers. Dr. Devoret's data (p. 107) clearly implicate repair by recombination mechanisms which, by sister exchanges of strands, allows for the dilution of those fragments containing thymine dimers.

Apirion: I assume that since there are strains that lack u.v. endonuclease but are still u.v. resistant, there should be at least one more system capable of excising u.v. products from DNA of irradiated cells.

Grossman: We now have sufficient data to indicate that our strain ML 7$^-$ is in fact a double mutant which besides lacking the endonuclease also lacks another gene—*X*. It is possible to segregate these markers by transformation, which allowed Dr. Mahler to list the following characteristics for *M. luteus*:

Strain	*U.v. sensitivity*	*Endonuclease*	*X gene*	*Transform-ability*	*X-ray sensitivity*
Wild type	+	+	+	+	+
ML 7$^-$	−	−	−	Poor	−
ML 7$^+$	+	−	+	+	+
C	−	+	−	Poor	−

+ Sensitivity implies resistance to the agent.

Moreover, those mutants lacking *X* are also sensitive to X-irradiation. It is apparent from these findings that resistance or sensitivity to u.v. is controlled not by the excision system but rather by the *X* system which because of its effect on transformation is in all likelihood this organism's recombination system.

Auerbach: Could one explain "hot spots" by the effects of the neighbouring nucleotides on the efficiency with which these substrates are used?

Grossman: The quantum yield and chemical reactivity of monomeric nucleotides is invariably greater than of these nucleotides in polymer form. The quantum yield for cytosine hydration is 16 times greater for CMP-5 than it is for CMP in DNA. As a result the most sensitive regions may in turn allow for more frequent incorporation of such analogues.

Auerbach: Do you think mutations are produced via the nucleotide pool?

Grossman: Not exclusively. All I am suggesting is that when one irradiates a cell with u.v. there may be a greater potential for mutagenesis to originate at the nucleotide pool level.

Maaløe: According to your model, mutations caused by pool nucleotides, modified by u.v., could only affect a small segment of the genome in a given cell, because the pool nucleotides would be used up quickly and incorporated into the segment of the genome replicated soon after irradiation. In an experiment with good synchrony, one would therefore expect to find a correlation between cell age and map location of the mutants.

Auerbach: Are either or both of your enzymes affected by caffeine?

Grossman: As far as we can tell, neither enzyme is affected at concentrations ranging from 0·1 M to 10^{-6} M. Moreover, repair in this organism is insensitive to caffeine.

Rörsch: We also find that the enzymes are not inhibited by caffeine. The discrepancy between the inhibition of repair observed *in vivo* and the absence of inhibition *in vitro* can be explained if the caffeine has to be bound to the DNA to exhibit its action and the binding occurs *in vivo* only when an energy source is present (H. E. Kubitschek, personal communication).

Maaløe: Did you say that there is a serious difference between *in vitro* and *in vivo* conditions as far as transcription is concerned?

Grossman: No, we have no data directly bearing on this, although data from biological experiments indicate that the transitions from C→T or GC→AT *in vivo* are mirrored in the experiments *in vitro*. The translational steps I think provide a confusing picture. With RNA and DNA polymerases, as we have already described, effective reduction of uracil or thymine leads to a loss of recognition at the polymerase level. Reduction is accomplished either by sodium borohydride or by photoreduction and dimerization. The lack of recognition can be attributed, in part, to a loss of hydrogen bonding capabilities associated with either direct reduction or indirect but effective reduction of the 5,6-double bond. This is accomplished by the addition of the elements of water or another molecule of uracil or thymine as in dimerization.

We showed earlier that the photohydration of some of the uracil residues of polyuridylic acid led to changes unlike those in replication or transcription, in that the translational coding properties of this synthetic messenger RNA rather than directing polyphenylalanine synthesis resulted in the formation of polyphenylalanylserine (Grossman, 1962, 1963). This change implies a coding transition from uracil to cytosine. More recently Ottensmeyer and Whitmore (1968) demonstrated that triplets containing uracil hydrates behave as cytosine in their ribosomal binding with aminoacylated tRNAs. Similar experiments by Cerutti, Miles and Frazier (1962), using triplets containing 5,6-dihydrouracil residues, led to completely different effects. In these conditions such triplets fail to absorb to ribosomes in the presence of charged tRNAs. Moreover, these same authors showed that poly 5,6-dihydrouracil is not able to hydrogen bond to poly A.

The differences in recognition in the translational system and those seen with the polymerases imply that forces in addition to hydrogen bonding complementarity may be in operation in amino-acid-incorporating systems. Moreover, the difference between uracil residues reduced at the 5,6-double bond by hydrogen or water leads me to believe that the presence of the hydroxyl group at C-6 provides for some difference in direct hydrogen bonding not available for such bonding with 5,6-dihydrouracil. Perhaps as result of a "backside" recognition it may be possible for a uracil hydrate to bond with guanine as if it were cytosine:

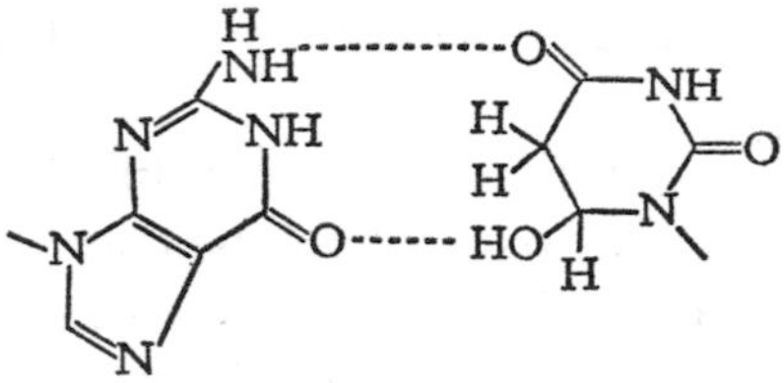

Guanine Uracil hydrate

Bridges: To explain the segregation of induced mutations or transformed genomes in bacteria, one may postulate that there is an enzyme which is capable of detecting a mismatched base pair. A dihydrocytosine must be about as near as one can go to a mismatched base pair. Is any enzyme known which will recognize this quite small distortion, which is a much smaller distortion than a pyrimidine dimer, for example?

Grossman: None that I know of. However, our working hypothesis is that the u.v. endonuclease is most likely anamorphous since excision or repair systems are not confined to u.v.-induced damage. The effects of a variety of other agents are repaired by excision systems in spite of the fact that their primary effects may be quite different. The u.v. endonuclease must, therefore, recognize the secondary effects which in all likelihood are distortions of the normal DNA conformation. Moreover, the u.v. endonuclease is a constitutive enzyme and is not induced by u.v. or other agents. The requirement for DNA and polydeoxynucleotides to act as substrates is currently under investigation in our laboratory.

Bridges: Have you tried the endonuclease on hydroxylamine-treated DNA?

Grossman: Not yet.

Evans: I believe Professor Rörsch has some mutants that are not u.v. sensitive, but nevertheless do not have the u.v. endonuclease. This would suggest that, at least in these cases, the u.v. endonuclease may not be involved in normal recombinational processes.

Rörsch: Repair occurs in Rec^+ Hcr^+ strains by recombination and by excision of dimers. Probably in *E. coli* the latter process deals with the majority of the u.v. lesions, whereas in *M. lysodeikticus* repair by recombination may be of greater importance.

Evans: I took your remark too literally, Professor Grossman. I understood you to say that this u.v. endonuclease may be a necessary requirement for recombination and indeed for dealing with conformational changes in normal DNA.

Grossman: The repair process requires at least four separate enzymic steps. At the same time recombination may require about eight to nine enzymes. A block to any one of these steps may impair one process without the other, or affect a step common to both. At this stage of our knowledge it is impossible to delineate the relationship between repair and recombination with excision.

Devoret: The deficiency in endonuclease I does not prevent recombination (Buttin and Wright, 1968).

Evans: In other words the mutants show recombination in the apparent absence of the endonuclease.

Devoret: Yes, and they are X-ray sensitive rather than u.v. sensitive.

Sobels: I am still a bit puzzled by the striking correspondence in Professor Grossman's data between the elimination of cross-links and of thymine dimers. Would these endonucleases recognize any kind of damage, or might there be different enzymes for removing cross-links and for excising thymine dimers?

Brookes: I would assume that the excision enzyme recognizes distortion of the DNA. I don't think any specific recognition is necessary.

Grossman: I agree. The excision system has a broad specificity. I hope that more details concerning the exact substrate specificity will soon be forthcoming.

Apirion: When those breaks are induced by either an enzymic or a non-enzymic reaction, would the next step be specific so that perhaps more than one enzyme excises all the different types of single-strand breaks?

Grossman: The plethora of nucleases present in an organism like *E. coli* provides for almost all conceivable kinds of incisions. The multiple nature of the enzymes and the multiple nature of activities associated with some nucleases (e.g. the DNA polymerase) seems to provide for all kinds of events.

Auerbach: Did you say that your exonuclease was not specific for u.v.-damaged DNA?

Grossman: Yes. The u.v. exonuclease exonucleolytically digests undamaged single-stranded DNA, u.v.-damaged single-stranded DNA and oligonucleotides as small as dithymidine monophosphate. Its lack of sensitivity to photoproducts is unique and as a result qualifies the enzyme to repair by excision through thymine photoproducts.

Apirion: Is there any reason to believe that there couldn't be an enzyme that recognizes the thymine dimer and excises it from both ends?

Grossman: I know of no photoproduct-specific nuclease.

Apirion: I suppose it would be interesting to know whether the first enzyme in the series you described starts by putting one nick somewhere near the thymine dimer, or two nicks from both sides of the same thymine dimer.

Grossman: Once it does make a nick on a double-stranded DNA it leads to a single-stranded region on which the enzyme, because of its specificity, doesn't act. The specificity of the enzyme is limited to native irradiated DNA. Since the dimers cannot hydrogen bond, a single phosphodiester bond break results in a small single-stranded region which cannot be acted upon by this enzyme. This region is now susceptible to the second enzyme, which is specific only for single-stranded DNA or regions of DNA (incised) which are single-stranded.

Auerbach: Is it known that this nick is made on one particular side of the photoproduct?

Grossman: The absolute mechanism of action of these two enzymes is now being examined. It is too early to indicate in exact terms just how they work.

Apirion: Why is a stretch of at least three thymines necessary in order to get a thymine dimer? Is it because a thymine dimer without a third thymine, on either side, would not be recognized and excised?

Grossman: I don't think this is so unusual. The available evidence is that thymine dimers and uracil-uracil dimers are preferably formed in pyrimidine-rich regions.

REFERENCES

BUTTIN, G., and WRIGHT, M. R. (1968). *Cold Spring Harb. Symp. quant. Biol.*, **33**, 259–270.
CERUTTI, P., MILES, H. T., and FRAZIER, J. (1962). *Biochem. biophys. Res. Commun.*, **22**, 466.
GROSSMAN, L. (1962). *Proc. natn. Acad. Sci. U.S.A.*, **48**, 1609.
GROSSMAN, L. (1963). *Proc. natn. Acad. Sci. U.S.A.*, **50**, 657.
GROSSMAN, L., and RODGERS, E. (1968). *Biochem. biophys. Res. Commun.*, **33**, 975.
JOHNS, H. E., LEBLANC, J. C., and FREEMAN, K. B. (1965). *J. molec. Biol.*, **13**, 849.
OKUBO, S., NAKAYAMA, H., SEKIGUCHI, M., and TAKAGI, Y. (1967). *Biochem. biophys. Res. Commun.*, **27**, 224–229.
ONO, J., WILSON, R. G., and GROSSMAN, L. (1965). *J. molec. Biol.*, **11**, 600.
OTTENSMEYER, F. P., and WHITMORE, G. F. (1968). *J. molec. Biol.*, **38**, 1–17.
RÖRSCH, A., SLUIS, C. A. VAN, and PUTTE, P. VAN DE (1966). *Radiat. Res.*, suppl. 6, 217.

NATURE OF ALKYLATION LESIONS AND THEIR REPAIR: SIGNIFICANCE FOR IDEAS ON MUTAGENESIS

P. Brookes, P. D. Lawley and S. Venitt

Chester Beatty Research Institute, Institute of Cancer Research: Royal Cancer Hospital, London

Very extensive use has been made of the alkylating agents in mutation studies since the classical discovery by Auerbach and Robson (1946) of the mutagenic action of mustard gas. This perhaps justifies a contribution to this symposium from an essentially chemical group having no experience of mutation induction or assay, but having an interest in the molecular biological mechanism of action of agents which alkylate cellular macromolecules.

After the initial work on the chemistry of alkylation of nucleic acids (reviewed by Lawley, 1966) it was necessary to test some of the ideas using a biological assay system. Since toxicity is the most easily measured biological property of the alkylating agents this has been studied most frequently, using bacteriophage (Brookes and Lawley, 1963), bacteria (Lawley and Brookes, 1965, 1968) or mammalian cells in tissue culture (Crathorn and Roberts, 1964). The results obtained were by no means the first in this field and the biological aspect has been much more extensively studied by others (e.g. Harold and Ziporin, 1958*a*, *b*; Loveless and Stock, 1959*a*, *b*, *c*, *d*; Loveless, 1959; see review by Loveless, 1966). The only new contribution made by the chemical approach was the use of radioactively labelled reagents which made it possible to replace the dose/time axis of the log survival curves by a more absolute value, namely the number of alkyl groups actually bound to the cellular macromolecules.

A very brief summary of our views which emerged from these studies would be that the cytotoxic and mutagenic action of the alkylating agents results from alkylation of the guanine moieties of DNA and that difunctional agents owe their very great toxicity to their ability to cross-link covalently the twin strands of the DNA double helix.

Since this symposium is concerned with the cellular processes leading to mutation but subsequent to the DNA alteration, we shall say nothing further on this primary step, our views having in any case been reported previously (Lawley and Brookes, 1961, 1963; Brookes and Lawley, 1964).

Since our contribution, if any, has been concerned largely with the quantitative aspect of alkylation it may be of value to discuss from this point of view three processes involving alkylated nucleic acids, which follow the modification of the DNA but must precede the appearance of the mutant cell, namely DNA repair, transcription, and messenger-RNA (mRNA) translation.

DNA REPAIR

The repair of otherwise lethal lesions in genetic material is clearly of significance in mutation studies. An effective mutagen must be able to introduce many potential mutagenic sites into the DNA but few lethal changes. This immediately raises the question of what constitutes a lethal lesion for any particular type of mutagen. For difunctional alkylating agents it appears highly probable that interstrand cross-links are responsible, by inhibiting the strand separation necessary for DNA replication. Hence for bacteriophages, viruses or micro-organisms which require total genome duplication for survival, one cross-link anywhere in the genome should prove lethal. The figure of two to three cross-links per genome found experimentally for T2 or T4 (Brookes and Lawley, 1963) and for *E. coli* B_{s-1} (Lawley and Brookes, 1968) is sufficiently close to one to suggest that this may well be the case. Therefore, for these organisms difunctional alkylating agents would be expected to be poor mutagens unless repair of these lesions occurred.

Largely as a result of the study of repair of u.v. damage in micro-organisms, it is presumed that four enzymic stages are involved, which may be summarized as excision, erosion (enlargement of the gap produced by excision), resynthesis and rejoining. After difunctional alkylation of *E. coli* B/r evidence for the first three of these processes was found (Lawley and Brookes, 1968), and selective removal of interstrand cross-links was demonstrated. As an example of the scale of this excision it was found that at the D_{37} dose of sulphur mustard for *E. coli* B/r there were 64 cross-links per genome of 9×10^6 nucleotides, while for the sensitive strain *E. coli* B_{s-1}, as mentioned above, two to three cross-links per genome constituted the lethal dose. From experiments in which the DNA had been pre-labelled with tritiated thymidine it was found that for each cross-link removed approximately 2000 nucleotides were also released. Since difunctional alkylation yields about five monofunctional alkylations for every cross-link, it follows that repair of the lesions could produce a viable cell containing many potentially mutagenic sites. Furthermore since the repair necessitates considerable breakdown and resynthesis, involving the monofunctionally alkylated DNA as template, it is possible that errors could be introduced into the DNA at this stage. Such a process might explain the observation of Alikhanian and Mkrtumian (1964) that, in actinophage, mutants resulted after difunctional alkylation which did not require phage replication to permit their detection. Freese and Freese (1966) also postulated a similar mechanism to explain the conversion of mutagen-treated T4 phage from mixed to pure mutant clone production.

The importance in mutation studies of the repair of lesions introduced by monofunctional alkylation does not seem to have been studied. In fact the nature of the cytotoxic lesion produced is not finally established, but recent evidence (Strauss *et al.*, 1966) seems to favour DNA strand breakage. Certainly DNA base alkylation itself is at best only weakly inactivating for T4 phage or *E. coli*, as evidenced by the extent of alkylation at the median lethal dose (Table I). Whatever the lesion produced, its repair is very efficient and for *E. coli* B/r at a D_{37} dose of methyl methanesulphonate or half sulphur mustard the level of cellular alkylation was as high as that achieved at a D_{37} dose of iodoacetamide, which was shown (Lawley and Brookes, 1968) to alkylate protein but not DNA or RNA. This suggests the possibility that inactivation in this repairing strain may be due to alkylation of cellular protein. The repair enzyme system itself is a possible target for inactivation, particularly as it was shown (Lawley and Brookes, 1965) that iodoacetamide could inhibit the excision enzyme in *E. coli* B/r. Loveless (1966) has already suggested that sensitization of irradiation damage in yeast by ethyl methanesulphonate may be due to alkylation of the repair enzymes.

TABLE I

INACTIVATION OF PHAGE AND BACTERIA BY ALKYLATION

Organism	*Nucleotides per genome* ($\times 10^6$)	*Extent of DNA alkylation at D_{37} inactivation dose** *Mustard gas*	*Half sulphur mustard*	*Methyl methanesulphonate*
T4 phage	0·36	12 (0·03)	400 (1·1)	600 (1·7)
E. coli B/r	9	320 (0·036)	17300 (1·9)	43500 (4·8)
E. coli B_{s-1}	9	43 (0·005)	790 (0·09)	15400 (1·7)

* Expressed as number of alkylations per genome and, in parentheses, as m-mole alkyl/mole DNA-phosphorus.

Clearly, it would appear from the above discussion that monofunctional alkylation is perhaps the ideal mutagenic event, since very many DNA bases can be modified per lethal event and even the lesions produced are readily repaired without loss of the potential mutational sites.

GENE TRANSCRIPTION

The stage of DNA transcription must precede the appearance of a mutant cell and so the effect of alkylation on this process was of some interest. The induction of β-galactosidase in *E. coli* B_{s-1} was studied after *in vivo* alkylation of the cellular DNA with di-(2-chloroethyl)-sulphide—mustard gas—or 2-chloroethyl 2-hydroxyethyl sulphide—half sulphur mustard (Venitt, Brookes and Lawley, 1968). It was shown that the mono- and difunctional agents had similar effects on inactivation of DNA as a template for mRNA synthesis and a level of reaction of approximately 1·4 m-mole/mole DNA-phosphorus was required to reduce survival of enzyme-synthesizing

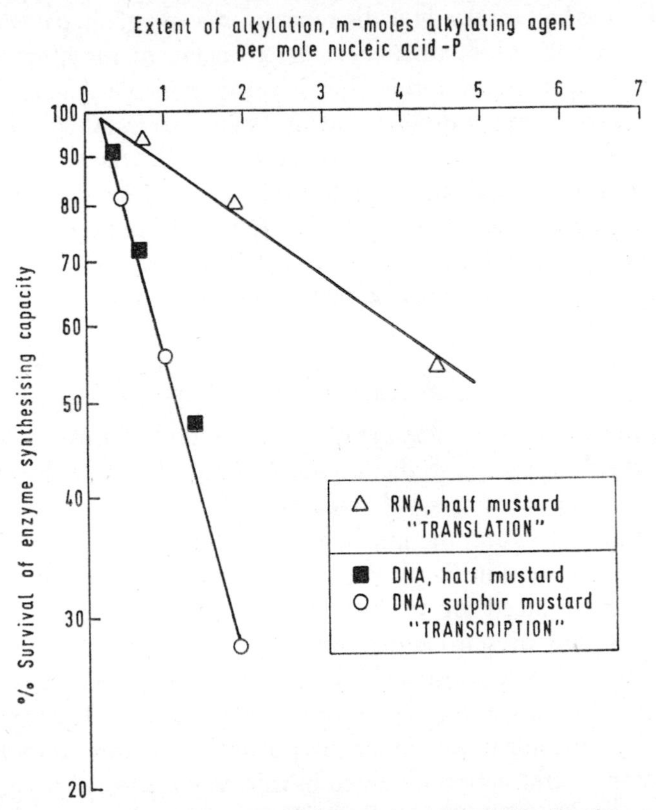

FIG. 1. Dose-survival curve of the capacity to synthesize β-galactosidase in *E. coli* B_{s-1} after alkylation by mustard gas or half sulphur mustard. For transcription the extent of alkylation refers to DNA, while for translation it relates to RNA alkylation.

capacity to 37 per cent of the control value (Fig. 1). If the polycistronic mRNA molecule for the *lac* operon containing the β-galactosidase messenger is assumed to contain approximately 3600 nucleotides (Kiho and Rich, 1964) then the corresponding DNA cistron would contain 7200 nucleotides. Since inactivation of DNA transcription requires a level of alkylation of approximately 1·4 m-mole/mole DNA-phosphorus, and assuming that only alkylation of the sense-strand of the DNA is significant, then five alkylations per cistron constitute a lethal hit. From the point of view of mutation studies it would be of interest to compare the level of DNA alkylation required to inhibit transcription with that resulting from a treatment designed to yield mutants. Unfortunately dose-response data for mutation by alkylating agents seem inadequate for this purpose but perhaps comparison with the D_{37} dose may be of some value. Table I shows that for the difunctional mustard the toxic dose is far below that affecting transcription, but for the half mustard the discrepancy is not so great, particularly for the repairing strain. Unfortunately we have no data on the effect of methylation of DNA on transcription, but it seems possible that with monofunctional agents, used at doses reducing survival to less than about 50 per cent, an effect on transcription may be expected. Since only enzyme activity was measured in the β-galactosidase induction experiments it is not known whether alkylation of the DNA resulted in non-production of the appropriate mRNA or whether an mRNA was produced containing base changes leading to production of an inactive enzyme.

MESSENGER RNA TRANSLATION

Since it has not proved possible to isolate individual mRNA molecules for any particular protein, studies on inactivation of mRNA by alkylation or any other means are necessarily inexact.

We attempted to get some information on this process, again using the β-galactosidase system in *E. coli* B_{s-1} (Venitt, Brookes and Lawley, 1968). The cells were induced for 30 seconds, then diluted 50-fold and mustard gas or half sulphur mustard added immediately. In this way it was hoped to alkylate the mRNA synthesized during the brief period of exposure to the inducer before it was translated into β-galactosidase. The rapid rate of reaction of the mustards used made this possible. The dose-response curve for inhibition of translation obtained in this way is shown in Fig. 1. The value for the extent of RNA alkylation was that determined for total cellular RNA and the assumption was made that the β-galactosidase messenger was similarly alkylated. The level of RNA reaction needed to reduce translation to 37 per cent of the control value was 6·8 m-moles/mole RNA-phosphorus. Uncertainties again arise regarding the size of the target involved in this inactivation. If it is that section of the mRNA coding for β-galactosidase and the enzyme subunit has a molecular weight of 40 000 (Weber, Lund and Wallenfels, 1963) then the cistron would contain approximately 1200 nucleotides and inactivation would result from about eight alkylations in this region.

Comparison of the data on inactivation of transcription and translation shows that the translation is about four to five times less sensitive to alkylation. The earlier discussion regarding the effect of mutagenic doses of monofunctional agents on transcription would apply also to effects on translation, but even higher extents of alkylation would be required to affect this process significantly.

In summary, the limited quantitative alkylation data available would suggest that difunctional agents would be too toxic to be effective mutagens in micro-organisms but would be most likely to produce gene deletion in organisms where this would not prove a lethal event. This would not apply to micro-organisms having an effective repair system which is envisaged to be very similar to that effecting dark-repair of u.v. damage. Studies of mutation frequency after difunctional alkylation of related organisms having *uvr*$^+$ and *uvr*$^-$ characteristics should indicate the significance of this process.

Monofunctional alkylation of DNA, particularly by methylating or ethylating agents, is relatively non-toxic, and at doses resulting in low survival these agents may affect cellular protein-synthesizing systems.

SUMMARY

A consideration of the quantitative aspect of the alkylation of cellular macromolecules suggests that for micro-organisms difunctional alkylation resulting in interstrand cross-linking might be too toxic to lead to effective mutation. However the operation of repair mechanisms may result in the expression of mutational events.

Monofunctional alkylation, however, would seem to be an ideal mutagenic event since many DNA base modifications are introduced per lethal event.

Quantitative studies of the effect of alkylation on transcription and translation in *E. coli* indicated that the use of monofunctional agents at moderately toxic doses might be expected to affect transcription, and perhaps with some agents even translation, which however was five times more resistant to alkylation than was transcription.

Acknowledgements

We thank Professor Sir Alexander Haddow for his interest in this work.

This investigation has been supported by grants to the Chester Beatty Research Institute (Institute of Cancer Research: Royal Cancer Hospital) from the Medical Research Council and the British Empire Cancer Campaign for Research, and by Public Health Service Research Grant No. CA-03188-08 from the National Cancer Institute, U.S. Public Health Service.

REFERENCES

ALIKHANIAN, S. I., and MKRTUMIAN, N. M. (1964). *Mutation Res.*, **1**, 355–361.
AUERBACH, C., and ROBSON, J. M. (1946). *Nature, Lond.*, **157**, 302–303.
BROOKES, P., and LAWLEY, P. D. (1963). *Biochem. J.*, **89**, 138–144.
BROOKES, P., and LAWLEY, P. D. (1964). *J. cell. comp. Physiol.*, **64**, Suppl. 1, 111–127.
CRATHORN, A. R., and ROBERTS, J. J. (1964). *Prog. Biochem. Pharmac.*, **1**, 320–323.
FREESE, E. B., and FREESE, E. (1966). *Genetics, Princeton*, **54**, 1055–1067.
HAROLD, F. M., and ZIPORIN, Z. Z. (1958*a*). *Biochim. biophys. Acta*, **28**, 482–491.
HAROLD, F. M., and ZIPORIN, Z. Z. (1958*b*). *Biochim. biophys. Acta*, **28**, 492–503.
KIIIO, Y., and RICH, A. (1964). *Proc. natn. Acad. Sci. U.S.A.*, **51**, 111–118.
LAWLEY, P. D. (1966). *Prog. Nucleic Acid Res.*, **5**, 89–131.
LAWLEY, P. D., and BROOKES, P. (1961). *Nature, Lond.*, **192**, 1081–1082.
LAWLEY, P. D., and BROOKES, P. (1963). *Biochem. J.*, **89**, 127–138.
LAWLEY, P. D., and BROOKES, P. (1965). *Nature, Lond.*, **206**, 480–483.
LAWLEY, P. D., and BROOKES, P. (1968). *Biochem. J.*, **109**, 433–447.
LOVELESS, A. (1959). *Proc. R. Soc. B*, **150**, 497–508.
LOVELESS, A. (1966). *Genetic and Allied Effects of Alkylating Agents*, p. 166. London: Butterworth.
LOVELESS, A., and STOCK, J. A. (1959*a*). *Proc. R. Soc. B*, **150**, 423–445.
LOVELESS, A., and STOCK, J. A. (1959*b*). *Proc. R. Soc. B*, **150**, 486–496.
LOVELESS, A., and STOCK, J. A. (1959*c*). *Proc. R. Soc. B*, **151**, 129–147.
LOVELESS, A., and STOCK, J. A. (1959*d*). *Proc. R. Soc. B*, **151**, 148–155.
STRAUSS, B., WAHL-SYNEK, R., REITER, H., and SEARASHI, T. (1966). In *The Physiology of Gene and Mutation Expression*, pp. 39–49, Prague: Academia.
VENITT, S., BROOKES, P., and LAWLEY, P. D. (1968). *Biochim. biophys. Acta*, **155**, 521–535.
WEBER, K., LUND, H., and WALLENFELS, K. (1963). *Angew. Chem.*, **2**, 481–485.

DISCUSSION

Maaløe: It is so delightful to get figures to work with that one wants to know precisely what they mean, and the experiment in which you tried to test the effect on translation seems to me to be ambiguous. By now we know something about the chain growth rate during transcription, and presumably it takes over a minute to go through the messenger stretch coding for galactosidase. The messengers that had begun to be synthesized towards the end of your induction period of 30 seconds will continue after the drug has been added.

Brookes: What you say is true but it is not easy to see how else the experiment could be done. We were mainly interested in seeing whether an effect could be observed and at what sort of dose levels.

Maaløe: Certainly, but my point is that messenger not yet synthesized could be modified by the drug both directly and via modifications in the DNA template. This makes the distinction between the effects on transcription and on translation uncertain.

Apirion: Perhaps similar experiments can be performed *in vitro* using phage RNA as a messenger?

Brookes: Inactivation of the RNA phage $\mu 2$ *in vitro* is being attempted at the moment, but results are not yet available.

Maaløe: Given that kind of result one could perhaps interpret the *in vivo* experiments with more assurance.

Evans: Your evidence clearly shows that five alkylations constitute the effective hit number whether you use difunctional or monofunctional compounds. But why should there be an equivalence in the efficiency of five monofunctional events and two and a half difunctional events in inhibiting transcription, unless a majority of the difunctional interactions are intrastrand?

Brookes: I don't see any problem. If the model for DNA with messenger RNA being synthesized on it is a three-strand intermediate of some sort, with the RNA being synthesized in the narrow groove of the DNA, then the fact that the DNA is cross-linked across the wider groove presents no difficulty.

Kubitschek: Whether the difunctional or the monofunctional agent was used, the complementary strand was also affected. In principle, if we can get down to the 20 per cent level and say these results were effectively the same with mono- and difunctional agents, we would have to say that even if the complementary strand were not directly affected, but was coding for translation, the monofunctional event acts just the same as a difunctional event. Therefore the unaffected strand can't be read under these circumstances. It doesn't matter whether the hit is on one strand or the other, on the face of these experiments.

Maaløe: Another point to consider here is that the process has to run to completion, whether it is transcription or translation, because you assay for completed molecules. Your figure of 5 for the relative sensitivity of the two processes seems to imply that on an average five point-modifications along a messenger molecule constitute "one hit" as far as translation is concerned.

Evans: What proportion of the difunctional products are intrastrand rather than interstrand?

Brookes: There is discussion on this point. Chemically it is simple: one hydrolyses the DNA and measures how much G is linked to G. Depending on the guanine content of the DNA this varies from 15 to 25 per cent. If there is 25 per cent G, then the chance of finding another guanine in the required position is one in four, and this roughly is about the amount of cross-link one finds, namely 25 per cent in a 50 per cent GC-containing DNA. From studies on denaturation and renaturation of alkylated DNA, G-G links measured chemically corresponded to interstrand links measured by renaturation ability (Lawley and Brookes, 1967). But then Dr. Lawley and Dr. Shooter looked into this in more detail, using phage T7. They thought this might be a better system with a somewhat smaller DNA, and easier therefore to isolate without single-strand breaks and so on. Some of the G-G links were definitely interstrand, but there was some evidence that perhaps some were not, so they were presumably intrastrand links. These results are still not too certain, since if the T7 DNA to start with had any single-strand breaks then renaturation studies become difficult to interpret quantitatively.

Rörsch: How many cross-links are introduced in single-stranded DNA?

Brookes: If one renatures mammalian DNA after denaturation it forms a tangled-up mess containing a certain amount of double-strandedness. However, with denatured DNA in 90 per cent methanol the strands stay apart, and under these conditions no G-G links are formed on treatment with mustard gas. This seems a good reason for believing that in single-stranded DNA no intrastrand linking can occur. It could be, however, that in 90 per cent methanol the base stacking is not the same as in water, so that adjacent guanines may not be in the same relationship to one another as guanines in DNA dissolved in water.

Rörsch: Did you compare the number of cross-links introduced in single-stranded φX 174 bacteriophage DNA and its double-stranded form (RF)?

Brookes: Some studies on this problem have been made by Dr. Loveless (cf. Loveless, 1966).

Kubitschek: How gross are the stereochemical changes produced by monofunctional agents? Presumably the enzyme must be tracking two-stranded DNA at the same time as it is producing RNA. So now it becomes of interest to ask how gross the physical or chemical change is in the DNA duplex after a monofunctional event.

Brookes: Of course one must introduce a change in the DNA since alkylation at N-7 quaternizes the nitrogen and introduces a positive charge. The positive charge could only be removed by splitting the bond between the purine and the deoxyribose, which occurs readily at 37° and pH 7.

A positive charge must introduce some disturbance in the molecule, but this is not detectable by physicochemical measurements as far as I know.

Kubitschek: Presumably even when the disturbance, whatever it is, is on the strand that is not being read for messenger, it is sufficient to inactivate translation.

Brookes: Why do you say that?

Kubitschek: Difunctional and monofunctional agents act with the same efficiency. One doesn't know where the hits are going to be, but the difunctional agent acts on both strands simultaneously, at least at the 20 per cent level. The monofunctional effects must be dispersed over both strands, although only one strand is coding for messenger.

Brookes: This translation-transcription story, of course, was raised from the point of view of events subsequent to DNA alteration; that is, whether effects on these processes were likely to occur with the levels of alkylating agents used to produce mutation. One would say certainly not for difunctional agents, but for monofunctional agents it is a distinct possibility.

Maaløe: You said that monofunctional agents may be more interesting as regards mutagenesis in micro-organisms than difunctional agents. You also said that perhaps there was a distinct difference between toxic and mutagenic events. Why do you think that "toxic" and premutational events differ except in their location within the genome?

Brookes: You are thinking in terms of a toxic mutation as opposed to a non-toxic mutation, but I am thinking more in terms of the chemical events.

Auerbach: In bacteria, one type of killing peculiar to the difunctional agents is cross-linkage. In higher organisms cross-linkage is one of the means, but not the only one, by which one can get chromosome breakage. Some years ago you wrote (Brookes and Lawley, 1964) that, *in vitro*, cross-linked DNA keeps on degrading without replication. This was the basis for experiments (Watson, 1964, 1966) in which we showed that when *Drosophila* spermatozoa are treated with a difunctional agent and then stored in untreated females, chromosome breakage goes on occurring for one or two weeks at a fairly high rate. At the end one may get 10 to 15 times as many breaks as at the beginning. But this does not happen with the monofunctional agents, although these can also cause chromosome breaks.

Brookes: That simply follows from the chemistry.

Clarke: The answer to that question may come from Dr. Kilbey's work. One could plot mutation frequency per survivor against percentage of survivors in the case of adenine-inositol reversions in *Neurospora*. Do the results for diepoxybutane, which is difunctional, differ from those with ethylene oxide, which is monofunctional? If they differ in the ratio of lethal to mutagenic events the curves shouldn't fall together.

Kilbey: Kølmark and I (1968) compared various aspects of their mutagenicity in the K3/17 strain I described in my paper (pp. 50–62). Only adenine reversions were studied. The following observations were made: (*a*) The mutagenic effect with both compounds was dependent on the total dose (time × molarity) over a wide range of molarities. (*b*) The mutation induction kinetics were similar. (*c*) The mutagenic efficiency is the same for both compounds. That is, at the same survival level both mutagens produce a mutagenic after-effect.

In *Neurospora* the two chemicals behave so similarly that, at present, there is no need to assume that DEB acts difunctionally either as a lethal or as a mutagenic agent.

Evans: Is it possible that the loss of an alkylated base after attack by a monofunctional agent could be the prime change that results in mutation? If this is so, then we should consider the kinetics of the depurination process and also of the subsequent chain scission. I believe that the half-life for depurination is around a day or two at 37°C, and chain scission is a very much slower process. Thus if depurination is the prime lesion that leads to mutation, then finalization of the mutation will be very much delayed in relation to the initial alkylation. Is there evidence for such a long lag or latent period of mutation induction with monofunctional compounds?

Brookes: There are two depurinations after alkylation: one is the loss of the 7-alkyl guanine and one is the loss of the 3-alkyl adenine. The 3-alkyl adenine is lost at about six times the rate of the 7-alkyl guanine, at 37°, pH 7. The amount of 3-alkyl adenine formed on alkylation of DNA is about one-tenth of the amount of 7-alkyl guanine, so roughly speaking after alkylation of DNA the depurination is equally likely to be either guanine or adenine. This depurination reaction is extremely sensitive to pH: if the pH drops one unit the rate may go up 10-fold so one has to keep the pH very constant. At 37° and pH 7 the half-life of depurination of DNA in a test tube is of the order of 40 hours or so. The depurination site will subsequently yield a strand break by a β-elimination process. Again the pH is critical. At pH 7 this is certainly a long process, with a half-life of perhaps 2000 hours. One would tend to think that single-strand breaks, introduced purely chemically, are possibly not significant. However, breaks could be introduced enzymically and for this to occur the depurination could be important.

Evans: From what you say, it should follow that mutations would be delayed in their appearance and indeed a majority would not appear until treated cells would have passed through a number of cell cycles after the initial alkylation. Surely there must be good genetic evidence either for or against this expectation?

Brookes: So your first question is, is there any evidence that depurination is mutagenic? That is a question to which I would certainly like to know the answer.

Clarke: Do Krieg's results (1963) with phage and ethyl methane-sulphonate support the concept of depurination? He got delayed mutations.

Loveless: I think all claims of an increase in the mutation frequency in T-even phages after monofunctional alkylation arising during a post-treatment storage period are without foundation. Professor Kaplan has evidence in phage κ that storage at low pH can result in mutation. I think chemically this must be attributed to depurination.

Kimball: Why do you say that these claims are without foundation, Dr Loveless?

Loveless: The original claim was made by Bautz, using T4 (Bautz and Freese, 1960). His increase was very small, but he attributed great significance to it. I found no evidence of it in my own studies (Loveless, 1959), and this was confirmed by Strauss (1961). In the course of many other experiments, there was no evidence whatsoever of an increase in T-even phage mutations during storage. All this work was with ethyl alkane-sulphonates, since until very recently these were the only alkylating agents known that would mutate these phages.

Devoret: If cross-links are excised the important repair process in the cell must be recombination between two nuclei, to rescue the information which is lost. Did you check whether one cross-link was lethal in *rec*$^-$ bacteria?

Brookes: We looked in detail only at *E. coli* B/r and B_{s-1}.

Kimball: Would you assume that the excision of many bases would go in one direction because of polarization along the strand? If it can go either way then information could be lost, but if it is polarized information would not be lost.

Brookes: If it is polarized there is still a cytosine opposite the guanine which is excised. The difficulty is how to stop the molecule falling apart. There is no reason why repair in each strand must go on simultaneously. It could be that one end is unhitched and repaired and then the other end could be excised. All we can say is that chemically we see not unhitching at one end only, which would still leave the guanine-guanine product attached to the DNA, but in fact its complete removal.

Auerbach: In higher organisms at a different level of organization chromosome fragments, presumably broken at the cross-link, can then join with each other and produce rearrangements, e.g. translocations.

Brown: I question whether the term depurination should be used so frequently in connexion with alkylation mutagenesis; there is no clear evidence that the spontaneous process is important. Depurination must remove information and the evidence from the T4 phage mutants that have been analysed suggests that one does not lose information about the base that has been affected in the mutagenic event. A fairly specific GC to AT type transition is indicated.

Loveless: When we come to an analysis of the nature of these mutations there is dispute. The Freese school interprets many ethyl ethanesulphonate mutants in T4 as transversions, but Krieg thinks the evidence is unsatisfactory. The simplest interpretation of the origin of transversion here is that it arises at an apurinic site.

Maaløe: If a cross-link region is excited, most likely the double strand is cut in two. Even without recombination between nuclei, this would only be a lethal event if it happened in the not yet replicated section of the genome. Thus many such events could happen to a uninucleate cell without loss of viability, even if there were no mechanism to join the two free ends.

Clarke: It has not been clearly established how many of the cross-links are interstrand as compared with intrastrand. One suggestion I have would be to treat DNA, isotopically labelled in one strand, with mustard, and degrade it. One should get three species: light G-mustard-heavy G, light G-mustard-light G, heavy G-mustard-heavy G.

Brown: Are you talking about the separation of macromolecules containing these residues or about the smaller fragments isolated after degradation?

Clarke: One would make a hybrid DNA molecule, treat it with the mustard, and then degrade it and see whether it is possible to separate these three species.

Brookes: We have tried this. The trouble is that the only way to separate these species is in a mass spectrometer. It can't be done with labelling. If one strand is labelled and the other strand is not, then one can't distinguish between labelled-unlabelled in a 100 per cent case, and fully labelled-fully unlabelled, 50 per cent of each. Dr Lawley hopes to do some work with the mass spectrometer, but the materials tend to contaminate the machine.

Clarke: If there is no real chemical evidence that there are lethal interstrand links then we have no problem about the excision repair.

Brookes: There is absolutely no doubt that interstrand links do occur on difunctional alkylation of DNA.

Clarke: But how many?

Brookes: One can show this very nicely by methods similar to those used by McGrath and Williams (1966). Density labelling is not needed. Difunctionally alkylated DNA is denatured in alkali, then centrifuged in an alkaline sucrose gradient. The alkylated DNA yields a species with a molecular weight double that of the controls (Venitt, 1968).

Clarke: But how many cross-links are needed to produce this effect? What is the ratio of interstrands to intrastrands? One cross-link between the single strands of DNA might be enough to hold them together.

Brookes: One would be enough within an unbroken length. The original estimates suggested that all cross-links were interstrand. More recent evidence suggests that this may not be the case, and maybe only one in three or one in four are interstrand—but certainly not less than that.

Apirion: Does the way that you measure the number of the cross-links distinguish whether they are interstrand or intrastrand?

Brookes: This is not really determined. It is easy to measure the total guanine linked to guanine moieties and that is the number we usually quote. The question that is difficult is where the two guanines come from.

Maaløe: The unit that you test for survival is the cell, with at least one and often two partly replicated units in it. How arc they all killed?

Brookes: It is probably relevant that one never finds in practice one cross-link per cell to be the mean lethal dose.

Maaløe: It is a little more complicated, because you have the age distribution of the replicating genomes to account for.

Loveless: It is perhaps significant that a u.v.-sensitive strain has a comparable relative sensitivity to a difunctional compound such as mustard gas, and nothing like the increment of sensitivity to monofunctional alkylating agents. And, *vice versa*, a strain which has been primarily characterized as mustard sensitive is also u.v. sensitive. If we consider just pyrimidine dimers in the one case and interstrand cross-linkage in the other, we are dealing with two lesions which, viewed anthropomorphically, are very different. It is difficult to see how a dark-repair system which has evolved in respect of u.v. lesions should find itself immediately capable of dealing with this very strange (and presumably non-existent in the evolutionary history of the organism) hazard of mustard gas cross-linkage. But if one looks at the intrastrand link it seems much more similar to a dimerization between adjacent pyrimidines—in fact it is an attachment of adjacent purines. If this isn't a totally unscientific way of arguing, one might consider that the recovery that we are observing is due to the removal of vicinal linkages rather than those across the dyad.

Brookes: There is definite evidence that this is not the case. One can show the loss of interstrand links during repair in *E. coli* by using the technique of alkaline gradient centrifugation. One can cross-link DNA in the cell, and during the process of repair in *E. coli* B/r one can see these cross-links being lost by the fact that the double molecular weight species disappears.

Clarke: But B/r has three repair systems and you don't know which one is effective. It may not be the excision repair system.

Kimball: Another point is that you showed that *E. coli* B_{S-1} was less resistant than the B/r to the monofunctional agent, so the B_{S-1} system can also repair with a monofunctional agent.

Brookes: Certainly, but we are not too sure what lesion is involved. I would say the alkylation *per se* is not the lesion, unless 14000 alkylations add up to a lesion.

Devoret: Is cross-linked phage DNA susceptible of being injected into the bacterial cell?

Evans: With a difunctional compound there is of course the possibility of a cross-linkage between the DNA and some molecule outside the DNA.

Kimball: Are you suggesting cross-linking of the proteins with the DNA?

Evans: Yes. We should not lose sight of the much-stressed fact that with a difunctional compound only some 20 per cent of the alkylations of DNA bases involve cross-linking within the DNA. The remaining 70 to 80 per cent of the difunctional molecules alkylate only a single guanine, so these are, in terms of the DNA, effectively monofunctional attacks. However, the other reactive arm of these molecules can interact with moieties surrounding the DNA. For example, is there any cross-linking to the messenger RNA that is closely associated with DNA and, particularly in the case of the eukaryotes, with the histone complex or other protein moieties on the chromosome? Have you any more recent evidence on these points, Dr Brookes?

Brookes: Only that we have never been able to find evidence for cross-linking between DNA and protein. First we looked simply to see whether after alkylation—the cellular alkylation of *E. coli* for example—the DNA isolated had more protein associated with it than had DNA from non-treated cells. With levels of alkylation well beyond the sensible biological doses we couldn't find any evidence of this. Obviously the limits of the method were rather narrow. We then used a histidine-requiring mutant and grew it in medium containing [^{14}C]histidine, so that the protein was heavily labelled. We then alkylated with sulphur mustard to see whether any ^{14}C was associated with the DNA. Again we could find no evidence for this. This just puts a limit to it; it doesn't say that it doesn't occur.

Apirion: What are the conditions for the isolation of the DNA?

Brookes: We used standard methods of isolation with phenol, which would not break any covalent bonds between DNA and protein.

Dawson: Is there any evidence that repressed and unrepressed genes respond differently to difunctional alkylating agents?

Auerbach: I do not know of any.

Böhme: There is evidence suggesting that the REC system participates in the repair of lesions induced by monofunctional alkylating agents or by photodynamic treatment. For example, in contrast to mutants with defects in the excision type of repair, *rec*$^-$ mutants of *P. mirabilis* (Böhme, 1968) as well as of *E. coli* (Böhme and Geissler, 1968) are highly sensitive to monofunctional alkylation and to photodynamic inactivation.

Brookes: I would like to think that single-strand breaks are the cytotoxic lesions for monofunctional alkylation, but the evidence is not there. In single-stranded DNA phage φX 174 there seems to be no doubt that one depurination is lethal. In double-stranded phage T7 about six to eight depurinations give a lethal effect. From that point of view I think one would say that depurination adds up to lethality.

Bridges: It is very tempting, and I have succumbed to it myself, to think of recombination as being a repair mechanism. But all we really know is that strains which don't recombine are more sensitive than strains which do. And this could be merely because in these cells DNA lesions may be lethal, whether radiation-induced or produced during the process of recombination. This doesn't tell you that recombination is involved in repair. There is a small but real difference here.

Devoret: Luria (1947) discovered multiplicity reactivation and thereafter cross-reactivation in phage T4 (Luria, 1952). So at first recombination was considered also as a kind of repair. It is only recently that more specific processes have been defined. I don't see the point of distinguishing recombinational repair from other repair processes.

Auerbach: But this doesn't really mean that recombination *is* the repair process. It could also mean that one and the same step is involved in both repair and recombination.

REFERENCES

BAUTZ, E., and FREESE, E. (1960). *Proc. natn. Acad. Sci. U.S.A.*, **46,** 1585–1594.
BÖHME, H. (1968). *Mutation Res.*, **6,** 166–168.
BÖHME, H. and GEISSLER, E. (1968). *Molec. gen. Genet.*, **103,** 228–232.
BROOKES, P., and LAWLEY, P. D. (1964). *Br. med. Bull.*, **20,** 91.
KØLMARK, H. G., and KILBEY, B. J. (1968). *Molec. gen. Genet.*, **101,** 89.
KRIEG, D. R. (1963). *Genetics, Princeton*, **48,** 561–580.
LAWLEY, P. D., and BROOKES, P. (1967). *J. molec. Biol.*, **25,** 143–160.
LOVELESS, A. (1959). *Proc. R. Soc. B*, **150,** 497–508.
LOVELESS, A. (1966). *Genetic and Allied Effects of Alkylating Agents.* London: Butterworth.
LURIA, S. E. (1947). *Proc. natn. Acad. Sci. U.S.A.*, **33,** 253–264.
LURIA, S. E. (1952). *J. cell. comp. Physiol.*, **39,** suppl. 1, 119.
MCGRATH, R. A., and WILLIAMS, R. W. (1966). *Nature, Lond.*, **212,** 534–535.
STRAUSS, B. S. (1961). *Nature, Lond.*, **191,** 730–731.
VENITT, S. (1968). *Biochem. biophys. Res. Commun.*, **31,** 355–360.
WATSON, W. A. F. (1964). *Z. VererbLehre*, **95,** 374–378.
WATSON, W. A. F. (1966). *Mutation Res.*, **3,** 455.

THE EFFECT OF RIBOSOME ALTERATIONS ON RIBOSOME FUNCTION, AND ON EXPRESSION OF RIBOSOME AND NON-RIBOSOME MUTATIONS

DAVID APIRION AND DAVID SCHLESSINGER

Washington University School of Medicine, St. Louis, Missouri

As a component of the protein-synthesizing machinery, the ribosome stands in a unique relation to the expression of mutations: it participates in the decoding process and can therefore give rise to errors or informational suppression of codons (Gorini and Beckwith, 1966), but is itself also subject to mutational changes of a rather unusual type. Mutation in one of its components can result in effects both on other components of the ribosome, and indirectly, by its role in translation, on other components of the cell. The pleiotropy of mutations in ribosomes is therefore of two kinds: (*a*) affecting other parts of the ribosome; and (*b*) affecting informational suppression, phenotypic or genotypic, mediated either by the ribosome or by tRNA. While the structural basis for these pleiotropic effects is still in great part conjectural, we want to discuss some examples relevant to each one, and how they may relate to the stabilization of the genetic code.

PLEIOTROPY OF RIBOSOME MUTATIONS

Organelles can be constructed so that many components interact with one another. A mutation in one part of the ribosome may therefore exert strong pleiotropic effects on other components. We have reported evidence for extensive pleiotropy and interactions of this type (Apirion, 1967; Apirion and Schlessinger, 1968*a*, *b*, *d*; Apirion and Schlessinger, 1969); a summary and extensions of the work are reported here.

Ribosome mutants of *Escherichia coli* were isolated as being more resistant or more sensitive than the wild type to an antibiotic that blocks ribosome function. Among these were mutants very resistant to streptomycin (Brock, 1966); spectinomycin (Davies, Anderson and Davis, 1965); lincomycin (Apirion, 1967); erythromycin (Apirion, 1967); neomycin and kanamycin (Apirion and Schlessinger, 1968*b*); and mutants very sensitive to lincomycin and erythromycin (Apirion, 1967). At least four mutants isolated in respect of each antibiotic were then tested for their response to other antibiotics not used in their isolation. In replica plating tests on various antibiotics, mutants often showed increased resistance to levels of an antibiotic somewhat higher than the parental strain could withstand, while some mutants were sensitive to levels at which the parental strain grew. As can be seen in Table I, all the six classes show pleiotropic effects of this type, i.e. they are more or less resistant than the parental strain to one or more antibiotics not used in their selection.

TABLE I

PLEIOTROPIC EFFECTS OF RIBOSOME MUTATIONS

	Effect	
Mutation	*Increased resistance to*	*Increased sensitivity to*
ery	*	spc
lin	cam, ery, spc	*
lir	*	cam, spc
nek	spc, str	cam
spc	cam, lin	*
str	ery, kan	cam, spc

* A number of tests were conducted, but no evidence for pleiotropy was found.

Abbreviations for antibiotics: cam, chloramphenicol; ery, erythromycin; kan, kanamycin; lin, lincomycin; neo, neomycin; spc, spectinomycin; str, streptomycin.

Five individually isolated mutants of each class were compared to the parental strain by replication tests on rich medium plates containing various levels of antibiotics to see whether the drug resistance is greater or less than that of the parent. The levels used were the following, in μg/ml: cam, 4 and 50; ery, 200 and 1000; kan, 2 and 80; lin, 250 and 1500; neo, 10 and 150; spc, 30 and 200; str, 2 and 200. For further tests and results, see Table II in Apirion and Schlessinger (1968*d*).

Another aspect of ribosome mutations which might apply to other indispensable organelles is the almost total ineffectiveness of certain mutagens. For example, ICR 191.A, a mutagen that induces frame-shift-type mutations (Whitfield, Martin and Ames, 1966), does not efficiently produce ribosome mutations like resistance to streptomycin or spectinomycin (Silengo *et al.*, 1967). This of course would be expected for mutations in a vital protein of an indispensable organelle; shifting the reading frame would completely ruin the structure of the entire protein chain after that point, so that the organelle would almost certainly fail to form or function.

INTERACTIONS BETWEEN RIBOSOME MUTATIONS

In strains carrying pairs of known ribosome mutations, we investigated whether the phenotype of the doubly mutant strains is simply the sum of the characteristics of the two singly mutant parental strains. Extensive interactions were found (Table II); i.e. in many cases the combination of two ribosome mutations in one strain leads to unexpected effects.

TABLE II

INTERACTIONS OF PAIRS OF MUTATIONS AFFECTING RIBOSOME RESPONSE TO ANTIBIOTICS

	str	*spc*	*nek*	*lir*	*lin*
ery	6	6	6	1	
lin	6	2 5 6	2 3 4 6	1	
lir	5	2 3 5	2		
nek	2 3	2 3 6			
spc	2 3 6				

The numerals indicate where significant interaction has been observed, corresponding to one of six categories.

1. Genotype looked for, but not found: in a strain containing a mutation in one class, mutagenesis was applied to look for a second mutation in another class; two classes were never found (*lir lin* and *lir ery*).
2. Reduction in mutation frequency: a double mutant class looked for as in 1, but appeared with rare frequency compared to direct mutation of the parental wild type to the second class.
3. Masked phenotype: introduction of a second mutation reduces resistance conferred by a resident allele.
4. Enhanced resistance: introduction of a second mutation increases resistance conferred by a resident allele.
5. New resistance: introduction of a second mutation confers a new resistance not found when either mutation alone is presented.
6. Informational suppression (phenotypic or genotypic) regained: a second mutation restores suppression restricted by the first. Further data and references are given in Table V and in Apirion and Schlessinger (1968*d*, Table 3).

For example, among the observed interactions were restriction of viable mutational changes in other ribosome loci (*forbidden genotype*); modification of an existing ribosome phenotype (*masked genotype*); masking of informational suppression, or restoration of masked informational suppression, phenotypic or genotypic; increased resistance to an antibiotic to which one of the parents was already somewhat resistant; gain of resistance to an antibiotic to which both parents were sensitive, etc. (Many of these studies are presented in some detail in Apirion and Schlessinger, 1968*d*, 1969; see also Table V, p. 162.)

One of the most severe cases is the interaction of *spc* (spectinomycin resistance) and *nek* (neomycin-kanamycin resistance) alleles (Apirion and Schlessinger, 1969). In this case the phenotype conferred by a mutation to spectinomycin resistance (resistance to more than 15 000 μg of spectinomycin/ml) is completely abolished by introducing into the strain a second ribosome mutation (*nek*). This masking of the *spc* phenotype was also demonstrated *in vitro*. The mutation to spectinomycin resistance could be crossed out from such strains, proving that it had been masked and not lost (Apirion and Schlessinger, 1969).

PLEIOTROPIC EFFECTS AND THE RIBOSOME AS A GUARDIAN OF THE CODE

Perhaps the most intriguing feature of the genetic code is its universality—not *that* the code is nearly universal, but *why*. Why doesn't the code drift? What constraint has been put on it?

Some possibilities could be discussed: for example, certain components might only fit together in a certain way. However, our partial knowledge at present, far from giving a simple answer, seems to argue the wrong way. The studies of informational nonsense and mis-sense suppressors have demonstrated that significant code-word changes can occur in bacteria by the alteration of tRNAs (Capecchi and Gussin, 1965; Carbon, Berg and Yanofsky, 1966; Gupta and Khorana, 1966) or by genetic alterations of the ribosome (Apirion, 1966; Rosset and Gorini, 1969). Why then does the code not drift? Why do the same codons continue to specify the same amino acids throughout the phylogenetic tree?

We would like to present some evidence that the ribosomes are one of the forces that help to conserve the code, and then to specify how this might happen.

There is evidence that changes in the ribosome itself can lead to suppression of a mis-sense codon (Apirion, 1966), and other evidence is available to indicate that ribosomal changes can also lead to suppression of nonsense codons (various mutants can suppress all three nonsense codons [Rosset and Gorini, 1969] or only two of them [Phillips, Apirion and Schlessinger, 1969, and in preparation].) However, ribosome mutations of these kinds are rather rare, and in general changes in the ribosome tend to inhibit nonsense and mis-sense suppression by tRNAs. After giving some of the supporting evidence for this claim, based on the effects of antibiotic mutations, we shall present a suggestion for rationalizing these effects.

(*a*) *Restriction of nonsense suppression*

Several groups of workers have pointed out that mutations to streptomycin resistance can eliminate or severely reduce amber suppression by *suI* (Reale-Scafati, 1967) and *suII* (Kuwano, Ishikawa and Endo, 1968) suppressors. We have confirmed some of these findings in respect of suppression by *suII* of three amber T4 phage mutants, and have also studied the effect of other ribosome mutations on suppression by *suII*. The data suggest that, like mutations to streptomycin resistance, other ribosome mutations can abolish suppression by *suII*.

(*b*) *Restriction of phenotypic suppression (curing)*

In a case which we have examined exhaustively, phenotypic curing of the allele *arg*$_3$ by streptomycin was severely depressed by mutations to streptomycin resistance (Apirion and Schlessinger, 1967). Several other alleles showed a similar effect. Here we have extended our studies of the strain harbouring the *arg*$_3$ allele to determine which antibiotic ribosome mutations restrict meaningful miscoding (Table III). In general, mutations that confer resistance to antibiotics that affect the 30S ribosome (streptomycin, neomycin and kanamycin) reduce phenotypic suppression much more severely than mutations in respect of antibiotics that affect the 50S ribosome (erythromycin and lincomycin). Spectinomycin, which is a "30S antibiotic", is the exception: mutations that confer resistance to it do not abolish phenotypic suppression of *arg*$_3$. This suggests that the *spc* protein (Apirion and Schlessinger, 1968*c*; Sparling *et al.*, 1968) affects the ribosomal binding site of the tRNA that is involved in the phenotypic suppression of *arg*$_3$ less than do the *str* and *nek* proteins (Apirion and Schlessinger, 1968*c*; Silengo *et al.*, 1967; Traub and Nomura, 1968; Sparling *et al.*, 1968).

TABLE III

LOSS OF CURABILITY BY RIBOSOME MUTATIONS

Mutants isolated on	*NG mutagenesis*			*U.v. mutagenesis*		
	No. tested	*Growth on str*	*Growth on kan*	*No. tested*	*Growth on str*	*Growth on kan*
Streptomycin	175	0	172	99	0	98
Lincomycin	188	184	181	204	202	202
Spectinomycin	188	169	172	210	210	210
Erythromycin	187	181	180	163	160	161
Kanamycin	166	128	2	191	189	0
Neomycin	133	96	5	174	166	3
Azide	194	194	194	92	92	92

Strain N663 (*thi*$^-$ *pro*$^-$ *arg*$_3$$^-$), derived from strain AB774 (Apirion and Schlessinger, 1967), was treated with NG (nitrosoguanidine) or u.v. (ultraviolet irradiation; see Apirion and Schlessinger, 1968*b*), and samples were plated on broth agar containing the various drugs (in μg/ml: str, 200; lin, 1500, spc, 200; ery, 1000; kan, 30; neo, 100; azi, 250). Mutants resistant to azide were isolated as a control, to show that non-ribosome mutations do not give the same effects. Colonies were transferred in an ordered pattern to broth agar plates (33 to each, including two original untreated parental colonies), and incubated at 37°C for 1 to 2 days. The master plates were then replicated to the following series of plates: MM (minimal salts medium)+thi+pro; MM+thi+pro+arg; MM+thi+pro+arg+1·5 μg streptomycin/ml; MM+thi+pro+arg+1·5 μg kanamycin/ml; and the same medium as the last two plates, but devoid of arginine. The parental strain N663 would grow well after three days of incubation on all these plates except MM+thi+pro, since the arginine requirement is phenotypically suppressed by either str or kan. The replicated colonies were scored after the three days for their ability or inability to grow without arginine in presence of str or kan. The listed results are from eight experiments, each of which started from a different colony. That the growth without arginine is due to suppression was shown by the ability of these drugs to support continuous growth of strain N663 in liquid cultures devoid of arginine in successive transfer, with a 10 000-fold increase of the inoculated mass. At the end of each experiment, all of the cells sampled and tested in each case (120) could be shown still to require arginine for growth.

(c) *Abolition of mis-sense suppression*

The battery of mutations induced in ribosomes to confer drug resistance also restrict mis-sense suppression. The test systems we have used are the ones in which Carbon, Berg and Yanofsky (1966) and Gupta and Khorana (1966) demonstrated suppression by modified tRNAs of mutations in the tryptophan synthetase A protein. In the former case, glycine is substituted for arginine, and in the latter, cysteine for arginine. As Table IV shows, all the classes of ribosome mutations tested can abolish mis-sense suppression; again, the mutations in the 30s ribosome were more restrictive than those in the 50s particle. In this case, in contrast to the case of phenotypic suppression (section (*b*) above), mutations to spectinomycin resistance were, in some instances, as effective in restriction as were mutations to streptomycin resistance. This suggests that the spectinomycin protein is important for the binding of the tRNAs involved in the mis-sense suppression.

TABLE IV

LOSS OF MIS-SENSE SUPPRESSION BY INTRODUCTION OF RIBOSOME MUTATIONS

Mutants isolated on	N1042				N1044			
	NG		*U.v.*		*NG*		*U.v.*	
	No. analysed	*Growth on MM*	*No. analysed*	*Growth on MM*	*No. analysed*	*Growth on MM*	*No. analysed*	*Growth on MM*
Kanamycin	135	76	85	47	36	28	†	†
Neomycin	†	†	77	69	20	7	20	14
Streptomycin	138	94	146	140	181	177	125	123
Spectinomycin	131	123	149	142	113	101	31	31
Erythromycin	74	72	124	124	84	76	59	58
Lincomycin	78	78	152	150	61	59	62	60
Azide	25	25	20	20	30	30	40	40

† Not tested.

Strains N1042 (*try* A23 su_6) and N1044 (*try* A78 su_9) (kindly supplied by Dr. C. Yanofsky, Stanford University), were treated with NG or u.v. and then plated on broth casaminoacids agar containing various levels of antibiotics (see legend to Table III). The colonies were arrayed on master plates and replicated to plates containing MM (minimal salts medium) and MM+tryptophan to test for suppression of the try^- allele. The results were recorded after two to three days. Strain N1042 can spontaneously lose its ability to grow on MM if it is propagated in broth, and therefore all manipulations were done in MM. After mutagenesis, the cells were grown, to permit the expression of induced mutations, in MM supplemented with tryptophan; they were then plated out on MM plates including tryptophan and drugs in the following concentrations, in μg/ml: azi, 250; ery 1000; kan, 20; lin, 1500; neo, 80; spc, 100; str, 100.

Results were obtained from five different experiments, each starting from a different colony.

From strain A78 su_9^+ (Carbon, Berg and Yanofsky, 1966), two of the spectinomycin-resistant derivatives (N553 and N567) and three of the streptomycin-resistant derivatives (N577, N578, N581) that no longer grew in the absence of tryptophan were studied in some detail. All showed growth rates identical to A78 su_9^+ in broth or minimal medium supplemented with tryptophan, but did not grow on unsupplemented minimal medium. Starved for tryptophan, even in the presence of the gratuitous inducer indole-3-propionic acid (Morse, Baker and Yanofsky, 1968), the five derivatives all developed less than one-tenth the level of tryptophan synthetase observed in the parental strain. It is of special interest that growth (and thus normal coding) seems relatively unaffected; the ribosome change specifically inhibits decoding by the particular tRNAs that suppress mis-sense mutations in the tryptophan synthetase A protein (see section (*e*) below).

Further genetic analysis by P1-phage-mediated transduction provided additional evidence that the suppressor mutation was still present in at least some of these strains, and that it was the mutations to drug resistance that masked the suppressor. For example, in transduction experiments, strain N567 (spectinomycin-resistant) donated the capacity to grow without tryptophan (su^+) to strain A78 (su^-) at a normal frequency, even though it could not itself grow without tryptophan. When the spectinomycin-resistant allele of strain N567 was transduced into A78 su^+, all 67 transductants examined had lost the ability to grow without added tryptophan.

In an even more exigent transduction test, N567 was the recipient of a non-restricting streptomycin-resistant allele. Because of the close linkage of spectinomycin resistance and streptomycin resistance (Davies, Anderson and Davis, 1965), the recombination with the incoming *str* gene usually displaces the *spc* allele from the chromosome. As anticipated, 88 streptomycin-resistant transductants were now spectinomycin-sensitive and, with the loss of spectinomycin resistance, 87 had regained the capacity to grow on minimal medium unsupplemented with tryptophan; three transductants that retained spectinomycin resistance still could not grow on minimal medium. (Details of these experiments will be published elsewhere.)

Finally, additional evidence that the suppressor mutation was still present in the restricted strains (e.g. N567) could be provided for some by restoring the suppression with a second ribosomal mutation, as follows.

(*d*) *Restoration of masked suppression by a second ribosome mutation*

As documented in Table V, strains with restricted phenotypic suppression occasionally regain suppressibility when a second ribosome mutation is introduced. All these combinations are rather rare and allele-specific (Table V); however, an auxiliary change in the configuration of the ribosome can restore the relationship of elements required for suppression.

A comparable restoration of *suII* nonsense suppression by mutations to spectinomycin resistance, in strains in which it had been abolished by

TABLE V

RESTORATION OF PHENOTYPIC SUPPRESSION BY AN ADDITIONAL RIBOSOME MUTATION

Strain no.	Induced by	Relevant genotype	Pheno. supp. by		azide			kan			neo			spc			str			ery			lin		
			str	kan	no.	str	kan	no.	str	kan	no.	str	kan	no.	str	kan	no.	str	kan	no.	str	kan	no.	str	kan
N811	NG	*nek*	−+	−+	†									93	*	*	59	$\underline{\underline{57}}$	$\underline{3}$ $\underline{\underline{13}}$	†			93	*	*
N812	NG	*str*	−	+	†			93	*	$\underline{\underline{93}}$	†			93	*	$\underline{\underline{2}}$				93	*	*	93	*	*
N813	u.v.	*lin*	−+	−+	†			†			†			60	$\underline{\underline{43}}$	$\underline{\underline{39}}$	60	$\underline{\underline{60}}$	$\underline{\underline{6}}$	60	$\underline{\underline{39}}$	$\underline{\underline{31}}$			
N814	NG	*nek*	−+	−	†									56	$\underline{36}$	$\underline{20}$	†			†			58	*	$\underline{6}$
N815	NG	*spc*	−+	+	†			60	$\underline{4}$	$\underline{\underline{60}}$	60	} $\underline{2}$ / $\underline{\underline{6}}$	$\underline{\underline{60}}$				60	$\underline{\underline{11}}$	$\underline{\underline{9}}$	†			†		
N822	NG	*nek*	−	−	20	*	*							99	$\underline{1}$	$\underline{7}$	60	$\underline{1}$	*	60	*	*	†		
N824	NG	*nek*	+	−	25	*	*							49	$\underline{\underline{3}}$	$\underline{31}$	32	$\underline{\underline{31}}$	*	55	$\underline{\underline{4}}$	*	45	$\underline{\underline{1}}$	$\underline{6}$
N827	u.v.	*nek*	+	−	54	*	*							51	$\underline{\underline{2}}$	$\underline{14}$	†			52	*	$\underline{1}$	50	$\underline{\underline{1}}$	$\underline{2}$
N835	u.v.	*ery*	+−	+−	32	*	*	†			†			50	$\underline{\underline{4}}$	$\underline{\underline{4}}$	50	$\underline{\underline{50}}$	$\underline{\underline{4}}$				50	$\underline{\underline{4}}$	$\underline{\underline{11}}$

† not tested, since preliminary tests done by direct replication of resistant colonies to test for phenotypic suppression gave negative results.
* tested, but no change from the parental strain was observed.
= the no. of colonies doubly underlined lost phenotypic suppressibility compared to the parental strain.
— the no. of colonies singly underlined regained phenotypic suppressibility compared to the parental strain.

The nine strains tested (column 1) were isolated from strain N663 (Table III) after NG or u.v. mutagenesis as indicated (column 2). Their relevant genotype is shown in the third column. The extent to which each was phenotypically suppressed by streptomycin and kanamycin is designated in the fourth column (+ good growth; +− fair growth; −+ poor growth; − no growth). We specifically chose for these experiments strains that had lost their ability to be phenotypically suppressed to various degrees in respect of one or the other drug (str or kan). Cultures of each strain were mutagenized and plated in broth agar containing the various antibiotics. Mutant colonies evincing the introduction of each of the second resistant alleles (top line, from column 5 to right-hand margin) were then tested for possible attendant restoration or further loss of phenotypic suppressibility. For further details about concentrations of drugs and replication tests of colonies, see legend to Table III. We note that restoration of phenotypic suppression is rare, specific to the combinations of certain kinds of mutations, and probably allele-specific. In only one case, strain N822, did a strain that had completely lost its capacity to be phenotypically suppressed regain it when a second ribosome mutation was introduced. Four other strains (N818 *lin*, N819 *spc*, N821 *nek*, N828 *str*) which had completely lost the capacity to be phenotypically suppressed were also tested, but in no case was restoration of phenotypic suppressibility found.

mutations to streptomycin resistance, was found by Kuwano, Endo and Ohnishi (1969).

For mis-sense suppression as well, restriction by one ribosome mutation could, in a few cases, be reversed by a second ribosome mutation. For example, from one strain N554, a *nek* derivative of A78 *su*$^+$ (see above) that could no longer grow without added tryptophan, 200 independent *str* derivatives were isolated. In 27 of them, with the introduction of the *str* allele, the strains regained the capacity to grow without added tryptophan. A comparable result was observed with strain N567, in which a *spc* allele had restricted the mis-sense suppression (section (*c*) above). Of 200 *str* derivatives, 12 had regained the capacity to grow without added tryptophan. In a third strain, N568, restriction by an *ery* allele was reversed in eight of 200 *spc* derivatives.

(*e*) *A model for the restriction of code drift by ribosomes*

We have in the past attributed the general restriction of miscoding by streptomycin-resistant mutations (Apirion and Schlessinger, 1967) to a loss in the mutants of some flexibility of the wild-type ribosome (see also Gorini and Beckwith, 1966). The argument was based on the "obvious assumption that sRNA molecules will have certain common features, and that the ribosome will ensure that all sRNA molecules are presented to the mRNA in the same way" (Crick, 1966). For phenotypic suppression by streptomycin (Gorini and Kataja, 1964) or mis-sense suppression by ribosome mutations (Apirion, 1966) a distorted ribosome may easily be visualized as bringing into juxtaposition a tRNA anticodon and a codon to which it pairs only by mistake. It is not surprising that in these cases mutational alteration of the ribosome can eliminate the "mismatch", which must be less stable than the true match of the normal codon-anticodon interaction.

Not so easy to understand is the tendency of ribosome mutations to restrict nonsense suppression that can result from modified anticodons in tRNAs (Smith *et al.*, 1966; Osborn *et al.*, 1967; Landy *et al.*, 1967; Osborn and Person, 1968; Goodman *et al.*, 1968), for here the modified tRNAs are truly matched with the corresponding codons. However, the suppression observed is a kind of competitive replacement for the polypeptide chain termination that tends to occur at nonsense codons, and modification of the ribosome could lead to a preference for termination instead of suppression. It can also be imagined that other factors might influence the efficiency of suppression. For example, neighbouring codons might interact slightly differently with tRNA when they precede or follow a nonsense codon, and ribosome mutations could thereby influence the efficiency of suppression indirectly.

However, it is impossible to use such an explanation for the restriction of tRNA mis-sense suppression by ribosome mutations (section (*c*) above), for in these cases the codon-anticodon interaction would appear to be normal, both at the critical codon and at neighbouring codons. Until appropriate tests have been carried out in cell extracts, the results (section *c*) are only suggestive; but should it hold true that mis-sense coding is much more severely inhibited than is normal coding, the ribosome mutation would be restricting *one* normal codon-anticodon interaction without restricting others. The assumption that all tRNA molecules are presented to the mRNA in exactly the same way seems insufficient here, for it suggests that any profound restriction of function of one tRNA by the ribosome should affect the function of many or all others.

We suggest, as an additional assumption, that the suppressor tRNA, with its modified anticodon, binds to the ribosome somewhat differently than did its unmutated form. In other words, tRNA molecules have some features that are *not* in common, since their primary sequences are different. The codon-anticodon interaction dominates the specificity of the binding of tRNA to ribosome; but most of the binding energy comes from interactions of the rest of the molecule with the ribosome. The non-codon part of the tRNA must arrange the binding so that the amino acyl residue is at the right point on the ribosome, and there may be classes of tRNAs with specific binding sites on ribosomes. When the anticodon of a tRNA is changed, the tRNA would then tend to fit the ribosome in a different way, and rather worse. Many changes in the ribosome that distort it would tend to weaken the mRNA-tRNA ribosome complex, and the binding of suppressor tRNAs, already weaker than normal, would be more affected. In short, the restriction of mis-sense suppression would be a severe example of the pleiotropy caused by the interdependence of the parts of a ribosome.

It is important to note that this suggestion is not in conflict with the notion of "wobble" (Crick, 1966). Wobble indicates that if the first two nucleotide pairs of the codon-anticodon triplet are fixed in space by the accepted base-pairing rules, the third tRNA nucleotide can be shifted by up to several Ångstroms from the dyad axis of the others. What is being suggested here is that the solid angle made by the entire codon-anticodon region with the ribosome can vary for different tRNAs, and for a mutant anticodon compared to its unaltered form. From the known dimensions of tRNA, a small angle that produces a shift of 0·1 nm (1 Å) at the anticodon (smaller than the shifts proposed in the case of wobble) could move the amino acid at the 3′-end of the tRNA by as much as 0·7 nm (7 Å). Such a shift could almost certainly not be allowed, and the sensitivity of some suppressor tRNA interactions to ribosome mutations thereby becomes more understandable.

Many findings can be explained by this kind of model. Probably the most important are the restrictions of miscoding by changes in the ribosome. All the observed restrictions can be explained by such a mechanism, rather than by the special ones applied to each case above. (For mis-sense suppression we cannot see any obvious alternative explanation.)

We have already encountered cases in which different antibiotics that bind to ribosomes affect the translation of one homopolymer much more than another homopolymer (Apirion, 1967), a situation that was explained in terms of slightly different binding sites or modes of binding for the different specific tRNAs involved. The curious dependence of poly-A-directed lysine incorporation on the level of added lysyl tRNA (Davies, Gorini and Davis, 1965) can also be explained by a different binding affinity of lysyl tRNA for ribosomes, compared to other species of tRNA. In these cases, the charge of the amino acid and interactions with neighbouring tRNA molecules on the ribosome can obviously also be important.

Another instance that became evident during these studies was the very different effect of mutations in the *spc* protein on the abolition of phenotypic suppression and on the abolition of genotypic mis-sense suppression (compare Tables III and IV). This difference would be easily accounted for if the *spc* protein were to participate in or influence the binding site for the mis-sense suppressor tRNA more than the binding site for the tRNA that effects phenotypic suppression of the *arg*$_3$ mutation. Again, this implies directly that the binding sites for the two tRNAs are slightly different.

CONCLUSIONS

The biological implications of these results are fourfold. First, the results suggest that the ribosome is an interdependent system of elements in which the conformation of many components is critical for its function in protein synthesis. Second, they suggest that the expression of mutations in genes that participate in the biosynthesis of an organelle is complicated by extensive interactions among the various elements of the structure studied. (This of course complicates immensely the genetic analysis of any multicomponent structure in the cell.) Third, the data further emphasize the importance of the genetic background in a strain (i.e. whether or not it contains altered tRNA molecules or altered ribosomes) for the analysis of mutagenesis of certain codons, and they draw attention to the possibility of negative mutagen specificity that can arise from differential killing of mutants. Fourth, according to the data and model presented here, the ribosome—*because of its complexity*—would tend to be stable during evolution and to help guard the stability of the code: for any change in the organelle would tend to interact throughout its structure, so that changes in any one of a great many regions could be lethal; and even permitted changes would tend to throw off any accumulated suppressor mechanism because the resultant tRNA-ribosome configuration would be relatively unstable compared to the standard ones. In agreement with such an overview, it is certainly striking that the universality of the code is paralleled by the fact that many parts of the ribosome are remarkably stable during evolution—the base ratios of rRNA are almost the same in bacteria of widely different DNA base compositions (Belozersky and Spirin, 1960); functional ribosomes can be constructed from RNA of *B. subtilis* and proteins of *E. coli* (Nomura, Traub and Bechmann, 1968); and in spite of the larger size of ribosomes from eukaryotes, at least some tRNAs of bacteria and higher organisms can be freely interchanged in protein synthesis in extracts (von Ehrenstein and Lipmann, 1961).

SUMMARY

Ribosome mutations that affect cell response to antibiotics were used as a probe to reveal extensive interactions of the parts of the ribosome. The biological implications are fourfold: (1) fixation and expression of mutations in an organelle are complicated by interactions among its elements; (2) the ribosome consists of a highly interdependent set of elements in which many components are critical to the function of each; (3) altered ribosomes or tRNAs in a strain can drastically affect the analysis of mutagenesis at certain codons; and (4) *because of its complexity*, many parts of the ribosome tend to be conserved during evolution, and thereby help to conserve the code: for even when a change does not produce deleterious interactions and lethality, it will tend to throw off suppressor mechanisms, which we suggest work through relatively unstable tRNA-ribosome binding.

Acknowledgements

This work is supported by NIH research grants, GM-10447, HD-01956 and Training Grant 5-T01-A1-00257, and by American Cancer Society Grant P-477. The help of Mrs. C. Mayuga is highly appreciated.

REFERENCES

Apirion, D. (1966). *J. molec. Biol.*, **16**, 285–301.
Apirion, D. (1967). *J. molec. Biol.*, **30**, 255–275.
Apirion, D., and Schlessinger, D. (1967). *Proc. natn. Acad. Sci. U.S.A.*, **58**, 206–212.
Apirion, D., and Schlessinger, D. (1968*a*). *XII Int. Congr. Genet.*, Tokyo, **2**, 51–52.
Apirion, D., and Schlessinger, D. (1968*b*). *J. Bact.*, **96**, 768–776.
Apirion, D., and Schlessinger, D. (1968*c*). *J. Bact.*, **96**, 1431–1432.
Apirion, D., and Schlessinger, D. (1968*d*). *Jap. J. Genet.*, Suppl.
Apirion, D., and Schlessinger, D. (1969). *Proc. natn. Acad. Sci. U.S.A.*, in press.
Belozersky, A. N., and Spirin, A. S. (1960). In *The Nucleic Acids*, Vol. III, pp. 147–185, ed. Chargaff, E., and Davidson, J. N. London: Academic Press.
Brock, T. D. (1966). *Symp. Soc. gen. Microbiol.*, **16**, 131–168.
Capecchi, M., and Gussin, G. (1965). *Science*, **149**, 417–422.
Carbon, J., Berg, P., and Yanofsky, C. (1966). *Proc. natn. Acad. Sci. U.S.A.*, **56**, 764–771.
Crick, F. H. C. (1966). *J. molec. Biol.*, **19**, 548–555.
Davies, J., Anderson, P., and Davis, B. D. (1965). *Science*, **149**, 1096–1098.
Davies, J., Gorini, L., and Davis, B. D. (1965). *Molec. Pharmac.*, **1**, 93–106.
Ehrenstein, G. von, and Lipmann, F. (1961). *Proc. natn. Acad. Sci. U.S.A.*, **47**, 941–950.
Goodman, H. M., Abelson, J., Landy, A., Brenner, S., and Smith, J. D. (1968). *Nature, Lond.*, **217**, 1019–1024.
Gorini, L., and Beckwith, J. R. (1966). *A. Rev. Microbiol.*, **20**, 401–422.
Gorini, L., and Kataja, E. (1964). *Proc. natn. Acad. Sci. U.S.A.*, **51**, 487–493.
Gupta, N. K., and Khorana, G. (1966). *Proc. natn. Acad. Sci. U.S.A.*, **56**, 772–779.
Kuwano, M., Endo, H., and Ohnishi, Y. (1969). *J. Bact.*, **97**, 940–943.
Kuwano, M., Ishikawa, M., and Endo, H. (1968). *J. molec. Biol.*, **33**, 513–516.
Landy, A., Abelson, J., Goodman, H. M., and Smith, J. D. (1967). *J. molec. Biol.*, **29**, 457–471.
Morse, D. E., Baker, R. F., and Yanofsky, C. (1968). *Proc. natn. Acad. Sci. U.S.A.*, **60**, 1428–1435.
Nomura, M., Traub, P., and Bechmann, H. (1968). *Nature, Lond.*, **219**, 793–799.
Osborn, M., and Person, S. (1968). *Proc. natn. Acad. Sci. U.S.A.*, **60**, 1030–1037.
Osborn, M., Person, S., Phillips, S., and Funk, F. (1967). *J. molec. Biol.*, **26**, 437–447.
Phillips, S., Apirion, D., and Schlessinger, D. (1969). *Bact. Proc.*, 128–129.
Reale-Scafati, A. (1967). *Virology*, **32**, 543–552.
Rosset, R., and Gorini, L. (1969). *J. molec. Biol.*, **39**, 95–112.
Silengo, L., Schlessinger, D., Mangiarotti, G., and Apirion, D. (1967). *Mutation Res.*, **4**, 701–703.
Smith, J. D., Abelson, J. N., Clark, B. F., and Goodman, H. M. (1966). *Cold Spring Harb. Symp. quant. Biol.*, **31**, 479–486.
Sparling, P. F., Modolell, J., Takeda, Y., and Davis, B. D. (1968). *J. molec. Biol.*, **37**, 407–421.
Traub, P., and Nomura, M. (1968). *Science*, **160**, 198–199.
Whitfield, H. J., Jr., Martin, R. G., and Ames, B. N. (1966). *J. molec. Biol.*, **21**, 335–355.

DISCUSSION

Auerbach: Do strains respond in the same way to whatever combination of resistance you introduce?

Apirion: Strain response depends on the combination of mutations induced.

Auerbach: Would you not expect curability to depend on the type of transfer RNA involved?

Apirion: Yes, but in order to observe such a specificity I suppose a large number of mutants at different codons should be analysed. However I do not believe that finding a statistically significant difference would tell us much more about the mechanism.

Auerbach: In the "curability" of auxotrophs there is specificity, isn't there? The fit must be spoiled specifically for certain messengers.

Apirion: Yes. The best evidence for this comes from testing streptomycin-induced miscoding, with alternating base copolymers as mRNAs (Davies, Jones and Khorana, 1966).

Auerbach: Are the two mutations which show most effect on translation, resistance to kanamycin and to streptomycin, both active on the 30S moiety?

Apirion: Yes.

Clarke: Do high, low or medium level streptomycin-resistant mutations, and streptomycin-dependent mutations, have similar effects on phenotypic suppression?

Apirion: Qualitatively they are the same. Quantitatively they are very different. In other words all levels of streptomycin resistance or dependence abolish phenotypic suppression levels of the *arg*$_3$ allele, but the frequencies with which it happens are very different (Apirion and Schlessinger, 1967).

Clarke: Is there evidence that transcription of these transfer RNA loci in fact occurs in the presence of the ribosomal mutations, Dr. Apirion? Are you sure that the ribosomal mutation isn't just stopping the production of the transfer RNA?

Apirion: At present this possibility is not excluded, and we are considering it, but it is very unlikely.

Magni: Dr. Clarke's question might be answered in the following way. In a strain carrying a mis-sense or a nonsense mutant and a suppressor of it one could introduce the resistance. If the suppressor is still active in the presence of the resistance, the hypothesis of block of sRNA would certainly be ruled out.

Apirion: Such experiments have been performed. However all they show is that the DNA for the *su*$^+$ tRNA exists intact in the restrictive cell. They do not answer the question of whether or not this DNA is transcribed and to what extent.

Auerbach: You said some resistance mutations affect the 30S moiety and others the 50S moiety of the ribosome. Does gene interaction depend on whether both resistance mutations affect the same moiety?

Apirion: Our limited data indicate that 30S–30S interactions are much more pronounced than 30S–50S or 50S–50S interactions.

Böhme: Do you get the same effect with regard to the masked phenotype when you combine the two mutants by recombination instead of selecting secondary mutations?

Apirion: Yes.

Kimball: If you start with spectinomycin resistance and select for neomycin resistance, then in general, from what you showed, the spectinomycin resistance decreases markedly, if it doesn't disappear entirely. If you go the other way, that is start with neomycin resistance and select for spectinomycin resistance, would you also expect the neomycin resistance to disappear?

Apirion: This is what happens.

Auerbach: But in either case you miss part of the mutations: those that are scored in the first type of experiment will be missed in the second.

Magni: When you have *spc* resistance and add a *nek* resistance, and then extract the *spc* alone by genetic analysis, this *spc* is still as resistant as it was before. Therefore is this only a phenotypic effect?

Apirion: Yes.

Magni: You mentioned that the half-mustard acridines are specific for frame shift. I gather that you don't assume that those few resistances you got were frame shift, but were a few mis-sense mutations among a large proportion of frame shift.

Apirion: We don't know whether or not the streptomycin-resistance mutations induced by ICR 191 were from the frame-shift type. However they differ from either spontaneously or nitrosoguanidine- or u.v.-induced streptomycin-resistance mutations in a very specific way. The difference is that, while all the ICR-induced mutations interact with a spectinomycin-resistance mutation to render the cell sensitive to spectinomycin, none of the other streptomycin-resistance mutations interacts with the spectinomycin-resistance mutation in such a way.

Magni: Are they different qualitatively?

Apirion: Yes. This was done in a specific strain where we could rigorously measure a forward mutation frequency for a dispensible protein. The experiments were carried out under conditions where ICR 191 was highly mutagenic (see Silengo *et al.*, 1967).

Magni: The proportion of mis-senses among the ICR-induced mutants is about 5–10 per cent. Could the amount you have found be that 5–10 per cent of mis-senses?

Apirion: Yes.

Kubitschek: Have you looked at the delay in expression of drug resistance? In our laboratory Dr. R. B. Webb has been looking at streptomycin resistance, using u.v. as a mutagen. It is an interesting finding that here the mutation seems to be expressed as quickly as he can observe it. Mutants accumulate immediately after u.v.

Clarke: This doesn't agree with the findings of Matney, Shankel and Wyss (1959), who used mutations to high level streptomycin resistance.

Apirion: Low level streptomycin resistance is not necessarily ribosomal, and therefore both observations are probably correct.

Kaplan: Several years ago we worked with low level streptomycin-resistant mutations (3 μg/ml) in a photoreactivation-negative strain of *E. coli*. We isolated about a dozen of these resistant mutations and tested their photoreactivability. Several of them had regained the potency of photoreactivation to different degrees. We couldn't explain this at that time, but now I think this can be understood as a case of suppression of the photoreactivation-negative mutation (restoration of the activity of the enzyme) by this second ribosomal mutation.

Auerbach: Is this true photoreactivation, not photoprotection?

Kaplan: This is photoreactivation.

Auerbach: Was the enzyme there?

Kaplan: I think there is enzyme again but it was not tested chemically.

Böhme: Does the restoration of photoreactivation occur in the presence or absence of streptomycin?

Kaplan: The mutants are streptomycin resistant and they show the restoration in its absence.

REFERENCES

APIRION, D., and SCHLESSINGER, D. (1967). *Proc. natn. Acad. Sci. U.S.A.*, **58,** 206–212.

DAVIES, J., JONES, D. S., and KHORANA, H. G. (1966). *J. molec. Biol.*, **18,** 48–57.

MATNEY, T. S., SHANKEL, D. M., and WYSS, O. (1959). *J. Bact.*, **78,** 378–383.

SILENGO, L., SCHLESSINGER, D., MANGIAROTTI, G., and APIRION, D. (1967). *Mutation Res.*, **4,** 701–703.

ASPECTS OF MODIFICATION OF NUCLEIC ACIDS IN MUTATIONAL PROCESSES

ADOLF WACKER AND PRAKASH CHANDRA

Institut für Therapeutische Biochemie der Universität Frankfurt am Main

CLASSICAL mutagenesis occurring through incorporation of base analogues into DNA, as for instance when bromouracil substitutes for thymine, is limited to altered DNA because thymine occurs specifically in DNA. Other chemical or physical agents do not show this specificity, because they change not only the DNA bases but also the bases in different RNA species. In this way they affect all cellular processes which influence genotypic expression.

BASE ANALOGUES

Replacement of the methyl group of thymine by a halogen atom leads to a significant shift in the electron density of the pyrimidine ring. For instance, bromouracil behaves like cytosine and is able to pair with guanine instead of adenine, as shown by Trautner, Swartz and Kornberg (1962) and Grunberg-Manago and Michelson (1964). Similar results can be obtained with chlorouracil and iodouracil which are also only incorporated into DNA. In contrast fluorouracil is only incorporated into RNA by pairing with guanine (Champe and Benzer, 1962). In this way only those processes are influenced which take place after DNA transcription. Mutagenesis does not occur.

2-AMINOPURINE

2-Aminopurine is another important compound for research on mutagenesis (Table I). Compared with bromouracil the specificity of aminopurine is marginal because it may behave like adenine or guanine and is incorporated into DNA as well as into RNA (Wacker, Kirschfeld and Träger, 1960). Like fluorouracil, aminopurine can inhibit the phenotypic expression of the genetic alteration. Our experiments with aminopurine-containing polynucleotides (Wacker *et al.*, 1966) have shown that aminopurine resembles adenine more closely than guanine (Wacker, Ishimoto and Chandra, 1967).

TABLE I

CODING PROPERTIES OF 2-AMINOPURINE POLYRIBONUCLEOTIDE (POLY AP)

mRNA	*Amino acid incorporation (counts/min per mg ribosomal protein)*		
	Lysine AA-Pu	*Glutamic acid GA-Pu*	*Arginine AG-Pu CGX*
−Polynucleotide	380	620	200
+Poly A	1992	796	180
+Poly AG (A:G 5·6:1)	4948	1432	820
+Poly AP	2648	1720	572

Conditions: polynucleotide concentration 120 μg/ml.
X=U, C, A, G.

RADIATION

Ultraviolet light

Radiation effects on DNA and RNA represent an important group of mutagenic reactions. U.v. irradiation of cells causes both mutation and death predominantly through thymine dimerization (Wacker, Dellweg and Jacherts, 1962). Besides dimerization thymine sustains other photoreactions of unknown biological significance (Dellweg and Wacker, 1962). Experiments with the thymine analogue 5-ethyluracil verify the importance of thymine dimerization in the lethal effect of u.v. irradiation. Incorporation of 5-ethyluracil into DNA desensitizes the cells to u.v. irradiation (Wacker, Gauri and Rüger, to be published).

TABLE II

EFFECT OF U.V. IRRADIATION ON POLYCYTIDYLIC ACID IN PROTEIN SYNTHESIS

U.v.dose	*Amino acid incorporation**			
erg/mm² × 10⁻⁵	*Proline CCX*	*Serine UCX*	*Leucine CUX UU-Pu*	*Phenylalanine UU-Py*
0	33	19	11·4	4·0
2·4	6·9	31	12·1	4·5
4·8	5·5	23	13·3	5·0
9·6	2·4	22	11·7	5·0

* Amino-acid incorporation as p-mole/mg of ribosomal protein in 20 minutes. Polycytidylic acid 320 μg/ml. Electrophoresis of u.v.-irradiated polycytidylic acid after alkaline hydrolysis shows ~6% UMP, u.v. dose $4 \cdot 8 \times 10^5$ erg/mm². X=U, C, A, G.

The mutagenic effect of u.v. irradiation is predominantly due to the deamination of cytosine to uracil (Wacker, 1963). After u.v. irradiation of polycytidylic acid (Table II) the incorporation of proline decreases sharply while that of serine and leucine increases significantly, the leucine less than serine (Wacker *et al.*, 1964). The coding triplets for both amino acids contain uracil in the first or second position. U.v.-irradiated cytosine residues in DNA yield other photoproducts of unknown biological significance (Dellweg and Wacker, 1962) (Fig. 1). Similar reactions take place with cytosine in RNA.

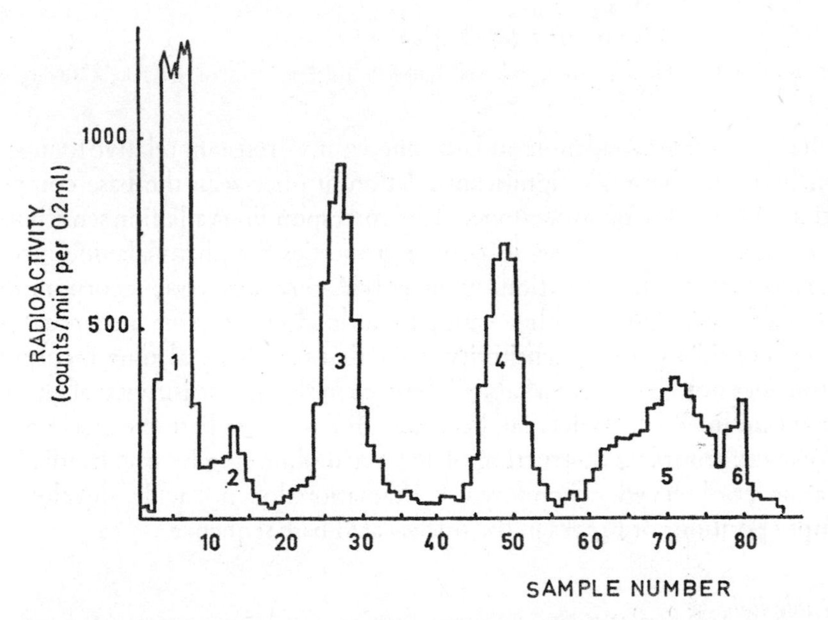

FIG. 1. Photoproducts of cytosine in DNA. DNA of *Streptococcus faecalis* R, labelled with [2-^{14}C]cytosine, was irradiated with u.v. (254 nm, 38×10^4 erg/mm^2). After hydrolysis (70 per cent perchloric acid, 60 min, 85°C) the material was chromatographed on a column (8×150 mm) of Dowex 2×8, 200/400 mesh, formate form. Elution was performed with a pH gradient of ammonium formate, and fractions (5 ml) were collected. Fraction 1: cytosine; 2: uracil dimer (mixed uracil thymine dimer ?); 3: uracil together with some thymine originating from the conversion of [2-^{14}C]uracil during growth of bacteria. Fractions 4, 5, and 6 not identified.

Under u.v. irradiation uracil sustains two main reactions: dimerization, and the addition of one molecule of water to the 5,6 double bond. Dimerization of uracil precedes biological inactivation, analogously with the DNA-inactivating effect of thymine dimerization. Addition of water, however, alters the electron density in the pyrimidine ring and possibly changes the coding properties of the base. Addition of water, therefore, could be considered as a mutagenic event. The inactivation of poly U by u.v. irradiation is shown in Table III (Wacker *et al.*, 1964).

TABLE III

U.V. SENSITIVITY OF NUCLEIC ACIDS IN PROTEIN SYNTHESIS

System	*Amino acid incorporation* (%)	
	Phenylalanine	*Lysine*
Complete	100	100
−mRNA	4	15
mRNA irradiated	6 (UU-Py)	92 (AA-Pu)
tRNA irradiated	89 (AAG)	78 (UU-?)
Ribosomes irradiated	92	102

Conditions: U.v. dose $2 \cdot 4 \times 10^5$ erg/mm²; polyadenylic acid 120 µg/ml; polyuridylic acid 80 µg/ml.

It can be seen that adenine and guanine are u.v.-resistant relative to uracil. Furthermore there is a significant relationship between the base composition of the codon or anticodon and the corresponding radiation sensitivity. For example, poly U loses its coding properties for phenylalanine incorporation after u.v. irradiation, while poly A-directed lysine incorporation is not affected. On the other hand, the anticodon for lysine incorporation shows considerable u.v. sensitivity while that for phenylalanine incorporation does not. This remarkable difference in the u.v. sensitivity of purine or pyrimidine-rich triplets can be seen in different irradiation experiments. However, complete destruction of the pyrimidine ring by u.v. irradiation was never observed. Therefore u.v. irradiation does not act by developing empty positions or breaks in the nucleic acid base sequence.

Visible light

Bromouracil incorporation into DNA of phages or bacterial cells sensitizes them to visible light. This effect is due to the small amount of ultraviolet in daylight or from daylight lamps (Wacker, Mennigmann and Szybalski, 1962). This small component is sufficient to convert bromouracil to uracil and to form further chemical derivatives of bromouracil, for example by reacting with cysteine (Smith and Alpin, 1966).

Table IV shows that visible light in the presence of peroxide or uranyl acetate reduces the u.v. photoeffect (Wacker *et al.*, 1964). Peroxide or uranyl acetate alone photoreactivate the u.v. damage to a lesser extent. Photoreactivation and monomerization of the uracil dimer are closely related effects. Irradiation experiments with pure polynucleotides can be used to interpret the changed properties of irradiated cells. Photoreactiva-

TABLE IV

PHOTOREACTIVATION OF U.V.-INACTIVATED POLYURIDYLIC ACID

Treatment of poly U	H_2O_2	*Uranyl acetate*	*Visible light*	*% Incorporation*
Non-irradiated	–	–	–	100
U.v.-irradiated				
$2 \cdot 4 \times 10^5$ erg/mm²	–	–	–	35
	+	–	–	76
	+	–	+	87
	–	+	–	73
	–	+	+	90

Conditions: Poly U was treated with hydrogen peroxide (3 μg/ml) or uranyl acetate (1 μg/ml) in the presence of visible light ($6 \cdot 6 \times 10^6$ erg/mm²).

tion of u.v.-irradiated cells increases the survival rate from 37 to 80 per cent (Table V).

The protective effect of peroxide could be due to a reaction with the 5,6 double bond of thymine and uracil which would inhibit dimerization. Besides this, there is some evidence that thymine dimers can be destroyed by peroxide in the presence of trace amounts of metal cations (Cu^{2+}, Co^{2+}, Fe^{3+}) and visible light or heat (Wacker *et al.*, 1964). This base destruction is comparable to the enzymic cleavage of thymine dimers from the DNA chain. This model may also be valid for the removal of uracil photoproducts from RNA. Fig. 2 demonstrates that thymine dimers are destroyed in the conditions described above.

In the presence of certain coloured or colourless compounds, visible light specifically degrades DNA and RNA purines. The dyes pyronin, thiopyronin (Wacker, Türck and Gerstenberger, 1963) and methylene blue (Simon and van Vunakis, 1962) initiate the photodynamic oxidation of

TABLE V

PHOTOREACTIVATION OF U.V.-INACTIVATED *E. coli* B (A) AND U.V. RESISTANCE OF *E. coli* 15T⁻ PRETREATED WITH HYDROGEN PEROXIDE (B)

System	*U.v. dose (erg/mm²)*	H_2O_2 *(3 μg/ml)*	*Visible light* ($\times 10^4$ *erg/mm²*)	*% Survival*
(A)	0	–	–	100
	200	–	–	37
	200	–	2	61
	200	+	2	80
(B)	0	–	–	100
	175	–	–	2
	175	+	–	3
	175	+	10	14

Conditions: In system (A) $5 \cdot 5$ ml of a bacterial suspension were u.v.-irradiated in a Petri dish, treated with hydrogen peroxide and reirradiated with visible light. In system (B) the bacterial suspension was first treated with hydrogen peroxide and visible light and then subjected to u.v. irradiation.

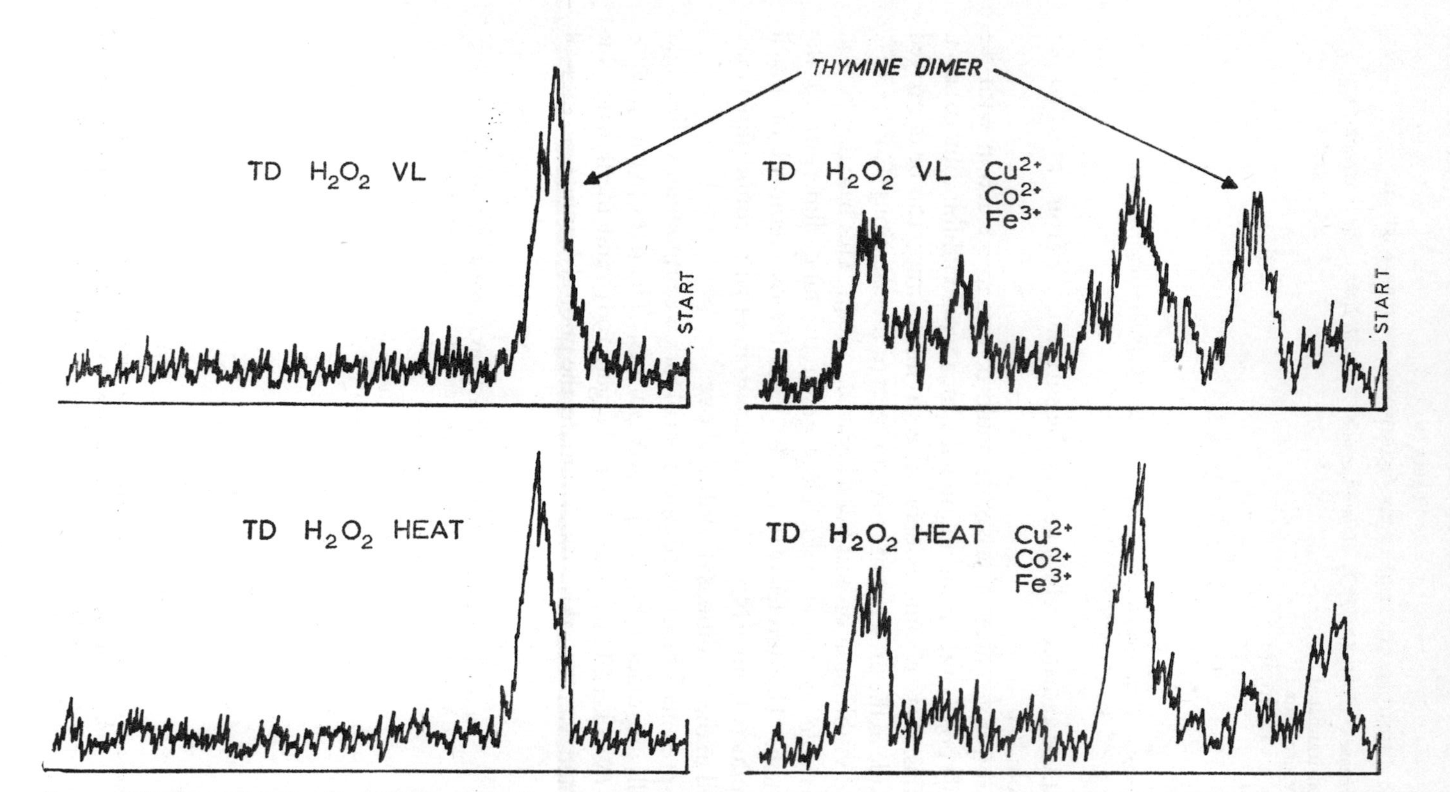

FIG. 2. Splitting of thymine dimer (TD) with hydrogen peroxide, metal ions, visible light (VL) or heat. [2-^{14}C]thymine dimer was dissolved in water and irradiated with visible light (1×10^6 erg/mm^2) or heated for 10 min at 90°C in the presence of hydrogen peroxide (3 mg/ml) and Cu^{2+}, Co^{2+}, and Fe^{3+} (3 μg/ml). Paper chromatogram: *n*-butanol/water, 86:14.

guanine in the presence of visible light and oxygen. Such a reaction may result in an empty position on the nucleic acid in place of a guanine. Upon replication a random replacement by other nucleic bases might occur. Therefore photodynamic inactivation should be similar to X-ray effects. Table VI shows that the photodynamic action of thiopyronin is closely associated with the alteration of guanine. The high sensitivity of ribosomes to photodynamic inactivation, relative to their normal behaviour under u.v. irradiation, should be noted.

TABLE VI

PHOTODYNAMIC EFFECT OF THIOPYRONIN ON DIFFERENT RNA FRACTIONS IN PROTEIN SYNTHESIS

	Amino acid incorporation (%)				
	Phenylalanine	*Lysine*		*Leucine*	*Proline*
System	*Poly U*	*Poly A*	*Poly AG**	*Poly UC**	*Poly C*
Non-treated					
Complete	100	100	100	100	100
− mRNA	4	15	15	3	3
Treated fractions					
mRNA	110	98	56	96	89
tRNA	94	96	96	78	20
Ribosomes	34	65			

Conditions: Fractions were treated with 10 μg thiopyronin/ml and irradiated with visible light ($6{\cdot}6 \times 10^6$ erg/mm^2).
* Poly AG (5·5:1); poly UC (4:1).

Many plants contain coumarins which react with thymine or uracil after visible light irradiation. For example, psoralen forms a photoadduct with thymine or uracil (Krauch, Krämer and Wacker, 1967) (Fig. 3). Irradiation in the presence of psoralen is largely lethal, but mutated cells (Fowlks, 1959) can also be observed. The psoralen photoadduct can be split into its components by u.v. irradiation, a reaction comparable to the monomerization of thymine dimers. These results contraindicate the use of psoralen in the treatment of vitiligo or for artificial tanning.

A very interesting photoreaction of thymine takes place under irradiation (at 315 nm) in the presence of aqueous acetone or similar ketones (von

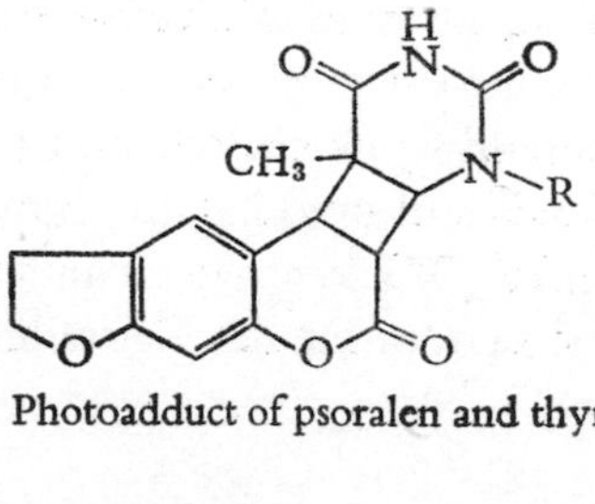

FIG. 3. Photoadduct of psoralen and thymine.

Wilucki, Matthäus and Krauch, 1967). In a similar fashion, with a low concentration of acetone, uracil yields uracil dimer exclusively, there being no addition reaction of acetone or water (Krauch *et al.*, 1967). The specificity of this reaction is reflected in the results in Table VII. Only poly U is changed; poly A and poly C remain unchanged (Wacker *et al.*, 1968).

TABLE VII

PHOTODYNAMIC ACTION OF ACETONE ON THE MESSENGER ACTIVITY OF POLYNUCLEOTIDES IN CELL-FREE PROTEIN SYNTHESIS

	Irradiated at 315 nm (3×10^5 *erg/mm*^{2}) (*% extinction*)		*Amino acid incorporation* (*counts/min per mg ribosomal protein*)		
Polynucleotide	*−Acetone*	*+Acetone*	*Lysine*	*Proline*	*Phenylalanine*
Poly A	100		1248		
		98	1368		
Poly C	100			389	
		97		401	
Poly U	100				5535
		43			794

Polynucleotides were irradiated at a concentration of 1 mg/ml containing 23% acetone.
The reaction mixture contained 80 μg poly U/ml, or 120 μg poly A or poly C/ml.

Although irradiation of bacteria under the conditions described is lethal, photoreactivation is more efficient than for eukaryotic cells. However, survival falls dramatically with repeated cycles of irradiation and photoreactivation. On the other hand bacterial survival remains high during repeated cycles of photoinactivation in the presence of acetone and photoreactivation. This difference may be due to certain irreversible photoreactions of thymine in the presence of u.v. light (H. D. Mennigmann, personal communication).

Irradiation with thermal neutrons and X-rays

Purine and pyrimidines are destroyed by irradiation with thermal neutrons in proportion to their nitrogen content. By a neutron dose of $0{\cdot}5 \times 10^{14}$ n, 9 per cent of thymine, 12 per cent uracil, 57 per cent guanine, and 75 per cent adenine are degraded (Wacker and Chandra, 1965). This is consistent with the altered biological activities of irradiated polynucleotides. As can be seen in Table VIII, adenine-containing code words are significantly more sensitive than those containing uracil, whereas the converse is observed after X-irradiation. Uracil and cytosine are, however, more sensitive to X-irradiation than adenine (Wacker and Chandra, 1964). Therefore the different sensitivities of the codons under X-irradiation may also be related to their base composition.

TABLE VIII

SENSITIVITY OF NUCLEIC ACIDS TO X-RAYS AND THERMAL NEUTRONS

	Amino acid incorporation (%)			
	Phenylalanine		*Lysine*	
System	*mRNA* (*UU-Py*)	*tRNA* (*AAG*)	*mRNA* (*AA-Pu*)	*tRNA* (*UU-?*)
Control	100	100	100	100
X-irradiated (80 krd)	29	86	69	77
Neutron-irradiated ($1\cdot8\times10^{13}$ n)	82	—	58	—

In general, different types of radiation affect cellular processes differently. Radiation-resistant codons on the DNA may be connected with radiation-sensitive codons on the mRNA. The corresponding anticodon again may be radiation-resistant. Generally speaking the DNA codons, the related mRNA codon and the tRNA anticodon show alternating radiation sensitivity (Table IX).

TABLE IX

ALTERNATION IN THE RADIATION SENSITIVITY OF BASES DURING TRANSCRIPTION AND TRANSLATION

	Transcription								*Translation*			
	DNA				*mRNA*				*tRNA*			
Radiation	*Purines*		*Pyrimidines*		*Pyrimidines*		*Purines*		*Purines*		*Pyrimidines*	
	A	G	T	C	U	C	A	G	A	G	U	C
U.v.			+	(+)	+	(+)					+	(+)
Visible light + TP		+						+		+		
Visible light + Acetone			+		+						+	
Neutrons (thermal)	+	(+)					+	(+)	+	(+)		
X-rays			+	+	+	+					+	+

+ = strong sensitivity; (+) = weak sensitivity. TP: thiopyronin.

CHEMICAL MUTAGENS

1-*Methyl-3-nitro-1-nitrosoguanidine* (*MNNG*)

Alkylating agents form a group of very active mutagenic compounds. In recent years we have concerned ourselves with one of these, MNNG, and in cooperation with Dr Lingens we investigated its mechanism of mutagenic action. We have shown that resistance towards MNNG and a diminished mutation rate are closely related to decreased MNNG uptake by the cells (Süssmuth *et al.*, 1969) (Table X).

TABLE X

ACTION OF MNNG ON VARIOUS CELLULAR PROCESSES IN *E. coli* B

System	Survival (%)	Mutation rate (%)	MNNG uptake (%)	Nucleic acid methylation (%)	
				RNA	DNA
Sensitive	2·0	11·7	100	100	100
Resistant (I)*	26·2	5·2	83	74	81
Resistant (II)*	53·7	3·7	50	58	57

Conditions: * Resistant (I) cells were obtained by a pretreatment of the sensitive strain with MNNG for 32 hours; further treatment of resistant (I) cells with MNNG for another 32 hours gave resistant cells (II). Incorporation and methylation experiments were done with [^{3}H]methyl-MNNG. The 100% level in controls was fixed arbitrarily.

Table XI shows that ribosomes are strongly inactivated by MNNG. Enzymes of the protein-synthesizing system, however, are not influenced.

TABLE XI

in vivo ACTION OF MNNG ON THE ACTIVITY OF RIBOSOMES IN PROTEIN SYNTHESIS

Fractions from non-treated bacteria		*Fractions from treated bacteria*		
Ribosomes	*pH 5 enzyme*	*Ribosomes*	*pH 5 enzyme*	*Lysine incorporation (%)*
+	+	−	−	100
−	−	+	+	41
−	+	+	−	40
+	−	−	+	98

Conditions: The bacteria (*E. coli* B) were grown in a yeast-glucose medium containing 30 μg MNNG/ml. Bacteria were centrifuged and washed free of MNNG. Lysine incorporation was measured using poly A (120 μg/ml) as messenger.

The mutagenic action of MNNG depends on the pH, the optimum being 5·5–6. In alkali MNNG forms diazomethane; at low pH it forms nitrous acid. The chemical basis for mutagenesis by MNNG at pH 6 therefore remains unknown.

TABLE XII

CODING PROPERTIES OF POLYNUCLEOTIDES TREATED WITH MNNG

	Amino acid incorporation (counts/min/per mg ribosomal protein)			
Polynucleotide	*Phenylalanine UU-Py*	*Leucine CUX UU-Pu*	*Lysine AA-Pu*	*Glutamic acid GA-Pu*
Without Poly U	99	185		
Untreated Poly U	4067	352		
Treated Poly U	3400	464		
Without Poly A			376	1157
Untreated Poly A			3014	981
Treated Poly A			1011	1644

Conditions: Polynucleotides were treated with 30 μg MNNG/ml at pH 5·6 and dialysed for 3 hours to remove excess of MNNG. Concentrations: poly U 80 μg/ml and poly A 120 μg/ml.

Poly C and poly A treated with MNNG (Chandra *et al.*, 1967) show a significant loss of their proline-and lysine-incorporating activities. The poly U-directed leucine incorporation and poly A-directed glutamic acid incorporation were changed after MNNG application (Table XII). Thus, the biological activities and the coding properties of DNA and of the different RNA species are strongly influenced by MNNG.

DISCUSSION AND CONCLUSIONS

Studies on the mechanism of cell-free protein biosynthesis have enabled us to interpret the cellular reactions involved in mutation. The fundamental process of DNA base-pairing discovered by Watson and Crick is also valid for all subsequent reactions participating in the information transfer from DNA through to phenotypic development. Many agents which cause mutational alterations do not change DNA exclusively. Only the thymine analogues specifically change the DNA properties. All other agents, such as aminopurine, irradiation or MNNG, change DNA as well as RNA bases. Replacement of deoxyribose by ribose or the different structures of single- or double-chain nucleic acids have only minor significance. Ribosomal RNA exhibits special resistance to mutation, possibly due to the bound ribosomal proteins.

Numerous irradiation experiments show that there is a systematic difference between the radiation sensitivities of purines and pyrimidines. Only pyrimidines are changed by u.v. irradiation; visible light in the presence of thiopyronin attacks guanine only, while in the presence of acetone it affects thymine and uracil only. Similar results were obtained with irradiation in the presence of psoralen. X-irradiation, however, primarily degrades pyrimidines.

During information transfer from DNA to mRNA and tRNA purines and pyrimidines alternate in a regular way. Therefore, under suitable conditions mutational effects on DNA can be compensated by corresponding events in tRNA. For example, the degradation of the second base of the DNA code and of the tRNA anticodon, obtained by X-irradiation or by visible light in the presence of thiopyronin, could possibly be without mutational significance for the irradiated cell. [In such a case, repair of radiation damage can succeed only by the cellular processes of protein synthesis. This model may provide a possible explanation for the low mutagenic activity of certain agents compared to their high lethality.]

Photoreactivation of a thymine dimer by visible light or heat in the presence of peroxide and traces of metal ions could be of great importance.

Visible light, traces of metals and even peroxide are omnipresent. Therefore this repair mechanism possibly represents an old evolutionary event.

SUMMARY

Most mutagens change DNA bases as well as the bases of mRNA, tRNA, and ribosomal RNA. But thymine analogues are specifically incorporated into DNA. All non-specific alterations of nucleic acids influence phenotypic development.

The coding properties of a poly AP were studied. Aminopurine codes like adenine as well as guanine.

Base destruction by different types of radiation was studied by means of cell-free protein biosynthetic systems. Irradiation by u.v. or visible light in the presence of acetone or psoralen specifically inactivates thymine and uracil. In addition u.v. irradiation without sensitizer causes deamination of cytosine to uracil. Visible light in the presence of thiopyronin specifically oxidizes guanine. Thermal neutrons mostly destroy purines; X-irradiation causes pyrimidine destruction. The presence of codons containing inactivated or changed bases alters the pattern of amino-acid incorporation in polypeptides "growing" *in vitro*.

1-Methyl-3-nitro-1-nitrosoguanidine (MNNG) acts on DNA as well as on RNA bases, mainly causing cell death. After application of MNNG we studied amino-acid incorporation in a cell-free protein-synthesizing system. Incorporation of leucine and glutamic acid stimulated by polyuridylic acid or polyadenylic acid is significantly influenced by MNNG treatment.

Investigation of cell-free protein synthesis in the presence of mutagens or treated polynucleotides or nucleic acids may be an excellent model for elucidation of the molecular mechanism of mutagenesis.

Acknowledgements

We wish to thank the Deutsche Forschungsgemeinschaft and the Verband der Chemischen Industrie—Fonds der Chemie—for their support.

REFERENCES

CHAMPE, S. P., and BENZER, S. (1962). *Proc. natn. Acad. Sci. U.S.A.*, **48**, 532–546.

CHANDRA, P., WACKER, A., SÜSSMUTH, R., and LINGENS, F. (1967). *Z. Naturf.*, **22b**, 512–517.

DELLWEG, H., and WACKER, A. (1962). *Z. Naturf.*, **17b**, 827–834.

FOWLKS, W. L. (1959). *J. invest. Derm.*, **32**, 233–247.

GRUNBERG-MANAGO, M., and MICHELSON, A. M. (1964). *Biochim. biophys. Acta*, **80**, 431–440.

KRAUCH, C. H., KRÄMER, D. M., CHANDRA, P., MILDNER, P., FELLER, H., and WACKER, A. (1967). *Angew. Chem.*, **79**, 944–945.
KRAUCH, C. H., KRÄMER, D., and WACKER, A. (1967). *Photochem. Photobiol.*, **6**, 341–354.
SIMON, M. I., and VUNAKIS, H. VAN (1962). *J. molec. Biol.*, **4**, 488–499.
SMITH, K. C., and ALPIN, R. T. (1966). *Biochemistry, N.Y.*, **5**, 2125–2130.
SÜSSMUTH, R., CHANDRA, P., WACKER, A., and LINGENS, F. (1969). *Z. Naturf.*, in press.
TRAUTNER, T. A., SWARTZ, M. N., and KORNBERG, A. (1962). *Proc. natn. Acad. Sci. U.S.A.*, **48**, 449–455.
WACKER, A. (1963). *Prog. Nucl. Acid Res.*, **1**, 369–399.
WACKER, A., and CHANDRA, P. (1964). *Angew. Chem.*, **76**, 685.
WACKER, A., and CHANDRA, P. (1965). *Angew. Chem.*, **77**, 428–429.
WACKER, A., CHANDRA, P., MILDNER, P., and FELLER, H. (1968). *Biophysics*, **4**, 283–288.
WACKER, A., DELLWEG, H., and JACHERTS, D. (1962). *J. molec. Biol.*, **4**, 410–412.
WACKER, A., DELLWEG, H., TRÄGER, L., KORNHAUSER, A., LODEMANN, E., TÜRCK, G., SELZER, R., CHANDRA, P., and ISHIMOTO, M. (1964). *Photochem. Photobiol.*, **3**, 369–392.
WACKER, A., GAURI, K., and RÜGER, W. To be published.
WACKER, A., ISHIMOTO, M., and CHANDRA, P. (1967). *Z. Naturf.*, **22b**, 413–417.
WACKER, A., KIRSCHFELD, S., and TRÄGER, L. (1960). *J. molec. Biol.*, **2**, 241–242.
WACKER, A., LODEMANN, E., GAURI, K., and CHANDRA, P. (1966). *J. molec. Biol.*, **18**, 382–383.
WACKER, A., MENNIGMANN, H. D., and SZYBALSKI, W. (1962). *Nature, Lond.*, **196**, 685–686.
WACKER, A., TÜRCK, G., and GERSTENBERGER, A. (1963). *Naturwissenschaften*, **50**, 377.
WILUCKI, I. VON, MATTHÄUS, D., and KRAUCH, C. H. (1967). *Photochem. Photobiol.*, **6**, 497–500.

DISCUSSION

Apirion: Can one induce miscoding by treating *in vitro* other parts of the protein-synthesizing machinery such as activating enzymes etc.?

Wacker: As I mentioned, the coding specificity of poly C is altered upon u.v. irradiation. Irradiation of tRNA and ribosomes leads primarily to their inactivation. These effects are due to specific alteration or destruction of bases by radiation.

Kaplan: The cytosine products were induced in the DNA by u.v. in the *in vitro* experiments. Did you try with the photoenzyme to get reversion or reactivation of the cytosine dimer and the other u.v. products?

Wacker: No, we have done such experiments only with thymine dimers. We found that thymidine dimers on incubation with yeast extracts and irradiation for 1 hour with an Osram Leuchtstofflampe (20 watts) form about 10 per cent thymidine monomer. Such studies are difficult to carry out with cytosine dimers; firstly because of their instability, and secondly because of conversion to uracil.

Sobels: Could you explain the contrasting effects you get with fast neutrons and X-rays?

Wacker: Our experiments were done with thermal neutrons only. In this case purines are more labile than pyrimidines. Whereas the destructive action of X-rays is due to the formation of oxygen-containing free radicals, that of thermal neutrons is probably due to the nuclear reaction, ^{14}N (n, p) ^{14}C.

Auerbach: Can you specifically change the pattern of transfer RNAs in the cell without using chemical methods?

Wacker: Recent research has shown that the tRNA patterns for some amino acids are dependent on several factors. Our own experiments have shown that during enzyme induction and under different nutritional requirements the patterns of tRNA are dramatically changed. This holds true for at least two specific tRNA types, namely for leucine and serine.

Auerbach: If this were involved in the broth effect of suppressor mutations, one would expect that different suppressors would respond differently to the broth effect.

Magni: That is the case. You get a different type of pattern of suppressors in different media.

Wacker: So in future work we should change the medium or the pattern of the tRNA.

Brown: The distribution coefficients of transfer RNAs change markedly with slight changes in the degree of base modification (e.g. methylation). It seems likely that changes in the chromatographic profile with change in culture medium are in large part a function of the level of base modification. It would require a much deeper analysis to establish specific suppressor tRNA changes.

Apirion: How many transfer RNAs did you look at for a change in the pattern with the change in growth medium?

Wacker: Two or three. I would think the specific tRNAs for those amino acids which have four or more codons should be tested for.

Magni: I don't see how by changing the anticodon one can change the amount of a specific tRNA charged with amino acid.

Wacker: We use a mixture of various tRNAs charged with a specific labelled amino acid. The elution pattern of this mixture on columns reveals one or several radioactive peaks. Each of these peaks is a tRNA specific for this amino acid.

Apirion: Since leucine has six codons and at least two different tRNA molecules it is possible to imagine that under different growth conditions the ratios of the different tRNAs vary, which might cause the different patterns observed.

Wacker: The new type of tRNA in the pattern may be necessary for the repair process.

Auerbach: This is what super-suppressors do. A new species of transfer RNA adds an amino acid to a nonsense codon in the messenger, and if it is acceptable at this site, the message can be read.

Hütter: Professor Wacker only said that by changing the anticodon of a specific transfer RNA, which carries the same amino acid as it did before the mutation, this tRNA may run differently on the column.

Kaplan: Do different peaks represent tRNAs with different anticodons?

Wacker: Not necessarily.

Auerbach: Ottensmeyer and Whitmore (1968) obtained phenotypic suppression of an amber mutation in phage by u.v.-irradiated transfer RNA.

Maaløe: Do you extract the tRNA with phenol and then strip the charged tRNA of its amino acid?

Wacker: We extract tRNA by phenol extraction of a 105 000 ***g*** supernatant, and purify by repeated precipitation with isopropanol. This is stripped at pH 9·0, charged with an amino acid and fractionated on columns. Experiments carried out under identical conditions with tRNA preparations from bacteria grown in different broths show different patterns.

Auerbach: Has it not been found that the spectrum of tRNAs in the cell differs with different growth conditions?

Wacker: Yes. There are many examples of this.

Apirion: In our laboratory a modified isoleucyl-tRNA was found in anaerobically grown *E. coli* compared to aerobically grown *E. coli* (Kwan, Apirion and Schlessinger, 1968). In no studies published thus far (to the best of my knowledge) have extensive changes in tRNA patterns been found when, for example, growth conditions or malignant and non-malignant cells were compared. Also, in no case except after virus or phage infection was a newly synthesized tRNA demonstrated. Where changes were found they were attributed to altered existing tRNA molecules rather than to the synthesis of new tRNA molecules.

Auerbach: A minor change might already lead to phenotypic suppression.

REFERENCES

KWAN, C. N., APIRION, D., and SCHLESSINGER, D. (1968). *Biochemistry, N.Y.*, **7**, 427–433.
OTTENSMEYER, F. P., and WHITMORE, G. F. (1968). *J. molec. Biol.*, **38**, 17–24.

RELATIONSHIPS BETWEEN RECOMBINATION AND MUTATION

G. E. Magni and S. Sora

Istituto di Genetica, Università di Milano

Genetic information is transmitted from cell to cell either in a conservative fashion or with variations due to exchanges of material between non-identical DNA molecules. We call the two types of transmission "replicative" and "recombinational" cycles in order to extend to micro-organisms the classical concepts of mitosis and meiosis. Normal vegetative reproduction of bacteria and of other haploid organisms, binary fission of protozoa and mitotic cell division in higher living forms fall into the first category. This does not imply, of course, that recombination is completely avoided during replicative cycles, but the exchange phenomena are rare exceptions to the conservative transmission of genetic information. Meiosis in eukaryotes, normal reproduction in many bacteriophages, conjugation, transformation and transduction in bacteria are all different aspects of the recombinational cycle.

A fairly large body of evidence has been collected during recent years about qualitative and quantitative differences between spontaneous mutations arising during the "replicative" and "recombinational" cell cycles. The only valid comparisons are evidently those in which the rates of spontaneous appearance of the same mutation are analysed during the two cycles in the same organism.

QUANTITATIVE DIFFERENCES

Bacteria and Ascomycetes have provided the most convincing data on quantitative differences between the two cycles. In *B. subtilis* the frequency of auxotrophs isolated by Yoshikawa (1966, Tables I and II) from normally growing cultures varies from 2 to 8 in 10^4 cells, while the same types of mutants are much more frequent among transformants of the same strains, reaching values one order of magnitude higher ($1 \cdot 2$ to $6 \cdot 7 \times 10^{-3}$). The phenomenon is caused by the integration of the homologous DNA.

M. V. Polsinelli (personal communication) was able to prove that in *B. subtilis* forward mutations for resistance to certain drugs (sulphanilamide, *p*-fluorophenylalanine) increase during transformation when DNA integration occurs in loci very strictly linked to the genetic determinants for resistance and not in other regions of the bacterial chromosome. The experiments performed by Demerec (1962, 1963) on "selfing" in *Salmonella typhimurium* are too well known to be discussed in detail. Many mutants (40 per cent in a sample of 201) show a much higher reversion frequency when the bacterium is transduced with phages carrying a DNA fragment homologous with the mutated locus than in uninfected controls.

In yeast our results (Magni and von Borstel, 1962; Magni, 1963, 1964, 1968) clearly indicate a very significant increase of mutation rates from mitosis to meiosis. We called this phenomenon the "meiotic effect" and have tested it in many different systems in *Saccharomyces cerevisiae*. In Table I a few examples of our findings are given. Spontaneous forward mutations for auxotrophy can be isolated only from haploid strains since they are always recessive. We tried to investigate the mitotic rate of mutation on two haploids without any selection procedure, by simple replica plating and from the meiotic spores of the diploid derived from their cross. Many auxotrophs were isolated after meiosis but we failed completely to obtain any mutants from mitotic cells. The ratio between mitotic and meiotic rates (Me/Mi) is therefore set at the minimal value of 70, but it could eventually turn out to be much higher.

TABLE I

MEIOTIC EFFECT IN *Saccharomyces cerevisiae* MUTATION RATES DURING MITOSIS AND MEIOSIS

Mutation	*Mutation rates* *Mitosis*	*Meiosis*	*Me/Mi*
Prototrophy→auxotrophy	$<5\times10^{-6}$	$3\cdot5\times10^{-4}$	>70
thr4–1→*Thr*	$0\cdot32\times10^{-9}$	$9\cdot9\times10^{-9}$	30
hi1–1→*Hi*	$1\cdot65\times10^{-9}$	$2\cdot4\times10^{-8}$	14·5
s(ar4–17)→*S(ar4–17)**	$8\cdot2\times10^{-8}$	$1\cdot3\times10^{-6}$	16·3

* *s*→*S*= mutation from inactive nonsense suppressors to the alleles active on the mutant *ar4–17* of the ochre type.

Experiments dealing with dominant mutations, such as reversions from auxotrophy to prototrophy, can be carried out on the same diploid strain homozygous for a given requirement, and the real mutation rates (mutations/cell per generation) can be estimated with high precision during both replicative and recombinational cycles. This was done on a large series of mutants and some of them showed increased reversion of the same order of magnitude as indicated in the second and third rows of Table I.

The meiotic effect was also investigated in another system (Magni and Puglisi, 1966), i.e. for mutations from inactive to active nonsense suppressors. The procedure does not differ significantly from that of a reversion test, the only complication being that each revertant must be genetically analysed to determine whether the reversion is due to a true back mutation at the nonsense triplet or to the appearance of a suppressor. The meiotic effect is clearly positive on the suppressor mutations, as shown in the fourth row of Table I.

Similar results, although restricted to the reversion of a few mutants, have been obtained in other Ascomycetes. In *Ascobulus immersus* an ascospore pigmentation mutant shows a tenfold increase in reversion at meiosis (Paszewski and Surzycki, 1964); in *Neurospora crassa* a very strong meiotic effect was observed for some mutations at the isoleucine-valine locus of group 2 (Kiritani, 1962; Bausum and Wagner, 1965).

It is therefore possible to conclude that in most micro-organisms, when a direct comparison is possible, all forward mutations occur much more frequently during recombinational than in replicative cycles. The phenomenon seems to be constant and valid for any kind of loci, as it has always been seen for genetic determinants involved in amino acid, purine and pyrimidine biosynthesis, for genes affecting drug resistance (see Table II), and for nonsense suppressors. Clear evidence is now available that some mutations show an increase of reversion rates during recombinational cycles and some do not, and this point will be discussed in detail below (p. 190 *et seq.*).

QUALITATIVE DIFFERENCES

When the first model was proposed to account for the higher mutation rates during meiosis (see last section) it became very important to answer the question about the specificity of the meiotic effect. In other words it was necessary to know whether any molecular type of mutation could occur more frequently during recombinational cycles, or whether the phenomenon was affecting only or predominantly particular mutagenic mechanisms.

Some data obtained for quite different purposes provide an indication of a certain degree of specificity of recombinational mutagenesis. We must restrict our discussion to the major molecular categories of mutations, i.e. base substitutions (mis-sense and nonsense) versus base insertions or deletions (frame shifts), because the available evidence is not statistically sufficient for further discriminations, such as, for example, transitions versus transversions or "+" versus "−" sign mutants. The molecular type of mutation spontaneously occurring during mitotic or replicative divisions has been indirectly determined in a variety of organisms. The evaluated proportion of mis-sense mutants varies to a great extent according to the organism investigated, to the procedure of isolating mutants and to the test for classification. In *Schizosaccharomyces pombe* Loprieno and Bonatti (1968) concluded that the great majority of spontaneous mutants at the *ad*-6 and *ad*-7 loci are mis-sense. In *Salmonella* Kirchner (1960) found a lower proportion (about 25 per cent) of mutants recognizable as mis-sense base substitutions, and in *Neurospora* de Serres (1964) reached the opposite conclusion, that spontaneous mutations at the *ad*-3*A* and *ad*-3*B* loci are mainly insertions or deletions. Clear-cut evidence was provided by Freese (1963) that most spontaneous mutants at the *rII* region of phage T4 (occurring therefore during a recombinational cycle) resemble in their reversion pattern the acridine-induced ones rather than those produced by base analogues and are presumably frame shifts. The differences in the organism and in the procedures do not however allow any definite conclusion. Once again it must be emphasized that the only reliable results are those obtained for the same mutation in the same organism during the two cycles.

We have approached this problem from two different points of view, of which only one will be discussed here; evidence from other experiments will be provided in the final section. It is generally agreed that temperature-sensitive mutations are due to base substitutions of the mis-sense type; only exceptionally could a base insertion or deletion give rise to mutants that are not complete at any temperature. We therefore investigated the meiotic effect on a particular mutation in yeast, i.e. canavanine resistance, which is now known to be due to a block in an arginine permease. Each mutant obtained at 33°C from mitotic or meiotic cells was tested at the permissive (23°) and non-permissive (33°) temperature and was classified as thermosensitive when capable of growing in the presence of canavanine at 33°C but not at 23°C. Table II summarizes one of our experiments. In the first section it is shown that for this mutation too the meiotic effect holds true, with a ratio Me/Mi of 5·3. The mitotic mutation rate of the diploid 5507 was not directly estimated, because *can*^{r} is recessive, but it was calculated from the corresponding rates of the two haploids from which 5507 was derived. The validity of this procedure for calculating mutation rates for recessive mutants in a mitotic diploid has already been discussed (Magni, 1964). From the second section of Table II it appears that the frequency of thermosensitive mutants is much lower among the meiotic canavanine resistants than among those isolated from mitotic cells. The difference is highly significant. Assuming that the occurrence of thermosensitive mutants does not increase at all during meiosis one should expect among the meiotic resistants a thermosensitivity frequency of 3·7 per cent (=19·6÷5·3), which is not significantly different from the observed value of 4·5 per cent.

TABLE II

EFFECT OF MEIOSIS ON BASE SUBSTITUTION MUTATIONS IN *Saccharomyces cerevisiae*

(1) Mutation rates from canavanine sensitivity to canavanine resistance (can^r) ($\times 10^8$)

Strain	*Mitosis*	*Meiosis*	*Me/Mi*
S 288C (haploid)	2·3	—	
5300/4c (haploid)	3·4	—	
5507 (diploid)	(5·7)	28·6	5·3

(2) Frequencies of thermosensitive can^r mutants

Origin	*Independent can^r mutants tested* No.	*ts**	*% ts*
From S 288C	42	10	24
From 5300/4c	60	10	17
Total from haploids	102	20	19·6
From 5507 after meiosis†	223	10	4·5

* *ts*= thermosensitive can^r.
† after subtraction of can^r mutants pre-existing in the cells before meiosis.

These results support two conclusions:

(1) The average molecular types of mutations arising during replicative and recombinational cycles differ from each other.

(2) The specificity of meiotic (or recombinational) mutagenesis seems to be real and indicates a proportional decrease of mutations of the missense type, i.e. of base substitutions. Mutations of the base insertion or deletion group consequently might be proposed as a majority product of spontaneous mutagenesis during recombinational cycles.

MUTATION AND RECOMBINATION

The main question which can now be raised is whether mutations arising during a recombinational cycle are in any way associated with genetic exchange. The only possible approach, at least at the present stage of development of our experimental material, is to investigate the relationships between mutation at any given locus and recombination of outside markers. This was done in three different organisms with more or less similar results. In the region *rII* of phage T4 the reversion of two frame-shift mutants is associated with 40 per cent recombination of outside markers, while only 15 per cent recombination was found among the revertants of two other mutants of the base substitution type in the same gene (Strigini, 1965). The intriguing isoleucine-valine (*iv*) mutants of group 2 were widely investigated by Kiritani (1962) and Bausum and Wagner (1965) in *Neurospora crassa.* In some isoallelic crosses the iv^+ revertants occur at very high frequency (sometimes up to 1 to 10 per cent) and these "anomalous" prototrophs are constantly associated with recombination of outside markers; but the recombination itself is also aberrant as only one type of recombinant is found among prototrophs. Needless to say a much deeper search for the mechanisms of these abnormalities is needed before the findings can be rightly interpreted in terms of relationships between mutation and recombination.

In yeast the mutational events occurring during meiosis are much less dramatic and more closely comparable with the corresponding phenomena observed in phage. The mutation *hi*1-1 is particularly suited for our purpose. It shows a clear meiotic effect, the locus is well located on chromosome V with proper markers on both sides, and a peculiarity of the meiotic reversion allows a critical experiment on recombination. Other mutations showing a positive meiotic effect always revert at meiosis at much higher rates than in mitosis, with no exceptions and with minor quantitative variations between independent experiments. The mutation *hi*1-1 is in a way aberrant: sometimes it fails to revert at high frequency during meiosis. The reasons for this behaviour, which we observed at least four or five times in about 20 independent experiments, are not yet known, but we got the impression that it depends upon some peculiarities of the chromosomal region of the *Hi*1 locus rather than being a peculiar property of the specific mutant. Table III summarizes the experiments in which the number of revertants recovered was large enough to allow analysis of the outside markers (as the usual reversion rate is of the order of 10^{-9}, the amount of revertants is often too low for any further study). The first row of the table shows the normal recombination frequency between the markers *thr*3 and *ar*6 in a random spore analysis. Experiments II and III show that recombination increases from 20 per cent up to 64 and 71 per cent as reversion rates increase (Me/Mi ratios of 14·5 and 27·0 respectively). In one experiment (I) where no meiotic effect was observed (Me/Mi = 1·0), the recombination frequency was identical to that of the controls. It must be noted that in the three experiments meiosis and sporulation were absolutely normal. The increase in mutation rates during meiosis is therefore directly correlated to recombination in the region of mutation.

TABLE III

RELATIONSHIP BETWEEN RECOMBINATION AND REVERSION AT THE *Hi1* LOCUS IN *Sacch. cerevisiae*

thr3 — *hi1–1* — *ar6*

Thr3 — *Hi1–1* — *Ar6*

	Parental	Recombinant	
	thr–3 ar–6 / *Thr–3 Ar–6*	*thr–3 Ar–6* / *Thr–3 ar–6*	Rec† (%)
Non-revertant spores hi	121	31	20
Revertant spores Hi			
I Me/Mi* = 1·0	43	12	22
II Me/Mi* = 14·5	12	22	64
III Me/Mi* = 27·0	47	116	71

* Three independent experiments; for explanation see text.
† Rec = recombination.

INTERPRETATION AND EXPECTATIONS

When we began our research on yeast one of us (Magni, 1963) proposed the model of "unequal crossing-over" to explain the facts known at that time. Subsequent investigations and the data from other organisms provided better support for this model in recent years. During meiosis unequal exchanges between homologous chromosomal regions cause the loss or the insertion of one or a few bases giving rise to frame-shift mutations. With the same mechanism mutations of the insertion or deletion type can revert through a mutational event of the opposite sign which restores the reading frame. This does not imply that unequal exchanges are the only source of mutations during meiosis. The same mechanisms responsible for mutagenesis in replicative cycles should also operate during recombinational cycles; unequal crossing-over would simply provide a new source of mutation and account not for the overall mutability but for the increase in mutations.

Streisinger and co-workers (1966) have recently proposed a detailed molecular model which takes into consideration the mechanism of exchange between homologous DNA molecules proposed for phages. They suggested that gaps can occur in one of the two strands of DNA in a region of repeating base sequences; a mispairing will follow the formation of the gap and eventually new DNA synthesis fills the gaps, causing the loss or the insertion of one or more bases.

These models account satisfactorily for all the observations discussed in the previous sections: the increase in mutations during recombinational cycles, the change in the molecular pattern of mutations with the proportional decrease in base substitutions in favour of base insertions or deletions, and the increase in recombination frequency among recombinational mutants. They also allow some specific predictions: (1) most forward mutations arising during recombinational cycles are probably of the insertion/deletion type. Freese's observations (1963) are in perfect agreement with this expectation, and also our data have shown so far that spontaneous meiotic mutagenesis is the most efficient way of obtaining frame-shift mutants in yeast. (2) The meiotic effect is probably specific for the reversions of frame-shift mutations and should be completely absent in the reversion of base substitution mutations. Convincing evidence in the expected direction is being accumulated in our material.

SUMMARY

Spontaneous mutations can occur during "replicative" or "recombinational" cell cycles. Meiosis in eukaryotes, the normal reproduction of some bacteriophages, and conjugation, transformation and transduction in bacteria are the recombinational cycles considered. It is shown that, when comparison is possible in the same organism, mutation rates are always higher during recombinational than during replicative cycles. The molecular pattern of forward mutations changes during recombinational cycles, with a decrease in base substitutions and a proportional increase in insertion/deletion mutations. Evidence is provided that recombination is often associated with mutation in the same region. The model of unequal exchanges provides a satisfactory explanation of the available data and allows some predictions to be made about the molecular nature of mutations arising during recombination.

REFERENCES

Bausum, H. T., and Wagner, R. T. (1965). *Genetics, Princeton*, **51**, 815–830.
Demerec, M. (1962). *Proc. natn. Acad. Sci. U.S.A.*, **48**, 1696–1704.
Demerec, M. (1963). *Genetics, Princeton*, **48**, 1519–1531.
De Serres, F. J. (1964). *J. cell. comp. Physiol.*, **64**, Suppl. 1, 33–42.
Freese, E. (1963). In *Molecular Genetics*, pp. 207–269, ed. Bryson, V., and Vogel, H. J. New York: Academic Press.
Kirchner, C. E. J. (1960). *J. molec. Biol.*, **2**, 331–338.
Kiritani, K. (1962). *Jap. J. Genet.*, **37**, 42–56.
Loprieno, N., and Bonatti, S. (1968). *Atti Ass. genet. ital.*, **13**, 293–297.
Magni, G. E. (1963). *Proc. natn. Acad. Sci. U.S.A.*, **50**, 975–980.
Magni, G. E. (1964). *J. cell. comp. Physiol.*, **64**, Suppl. 1, 165–172.
Magni, G. E. (1968). *XII Int. Congr. Genet.*, Tokyo, **1**.
Magni, G. E., and Borstel, R. C. von. (1962). *Genetics, Princeton*, **47**, 1097–1108.
Magni, G. E., and Puglisi, P. P. (1966). *Cold Spring Harb. Symp. quant. Biol.*, **31**, 699–704.
Paszewski, A., and Surzycki, S. (1964). *Nature, Lond.*, **204**, 809.

STREISINGER, G., OKADA, Y., EMRICH, J., NEWTON, J., TSUGITA, A., TERZAGHI, E., and YNOUYE, M. (1966). *Cold Spring Harb. Symp. quant. Biol.*, **31**, 77–84.
STRIGINI, P. (1965). *Genetics, Princeton*, **52**, 759–776.
YOSHIKAWA, J. (1966). *Genetics, Princeton*, **54**, 1201–1214.

DISCUSSION

Auerbach: A large proportion of the spontaneous mutations which you obtained in meiosis should again show a strong meiotic effect on reversion.

Magni: That is right.

Kimball: Have you checked specifically that the mutants arising in meiosis were associated with recombination of the end markers or outside markers?

Magni: We have only isolated 50 meiotic mutants so far, and we don't apply any selection system. They are isolated at random and the frequency of the order of 10^{-4} makes it improbable that repeats would be found in the same gene.

The only possibility of checking this point would be to correlate the occurrence of canavanine resistance with recombination. Unfortunately nobody has been able to locate the gene for canavanine resistance. We are now testing other types of resistance due to genes whose position in the chromosome is known. We hope to be able to prove that forward mutations arising during meiosis are associated with outside marker recombination.

Apirion: I believe we should still be a little cautious in concluding that a frame-shift mutation results from a recombinational event. Perhaps the correlation observed between the two is coincidental, since both could be enhanced by the same event—for instance, repair. We should keep those possibilities in mind because it is difficult to explain by recombinational mechanisms how, for instance, acridine half-mustards lead to frame-shift mutations in vegetative bacteria.

Magni: This, on a reduced scale, is what has been known since 1930 for the *Bar* duplication in *Drosophila*.

Auerbach: No, this is misleading. In the case of *Bar* the duplication is necessary; it has to be there first. Now whether this is necessary in your work is not so sure. You may by accident get the matching of a few nucleotides in different parts of the gene.

Magni: The repeat sequence in the same gene is operationally a duplication.

Auerbach: The duplication does not necessarily have to be in tandem. It could be in a different position.

Clarke: Roland Megnet (1966) had some evidence in *Schizosaccharomyces* that 5-fluorouracil deoxyriboside greatly stimulates recombination. Have you used this in *Saccharomyces*?

Magni: We tried to use fluorouracil, as 5-fluorouracil deoxyriboside is not taken up by *Saccharomyces cerevisiae*. Unfortunately fluorouracil completely inhibits sporulation and the experiment cannot be done properly. Now we are trying different temperatures. Variations of temperature do not affect the reversion rate of our mutants, at least not significantly. We run the experiments at 13 and 28°C. Sporulation and meiosis occur very well within this range of temperature and the recombination frequency varies significantly, being higher at 13° than at the standard temperature of 28°C. The correlation between mutation rates and recombination frequencies is going to be tested.

Hütter: Would temperature have an important effect on the fixation of mutation?

Magni: I don't believe so. Fixation should occur much later, at the time of spore germination which can take place even days after sporulation, and always at the same temperature (28°C).

Hütter: Did you find a temperature effect also during the replicative cycles? Dr. Zimmermann in Freiburg gets quite different mutation frequencies at 25° and 35°C (Zimmermann, Schwaier and von Laer, 1966).

Magni: The mutants we have so far tested (very few indeed) do not show any change of mutation rate at low temperature during mitosis. It is important in this respect to know the molecular nature of the mutations analysed. It could be that different types of mutants react differently to temperature variations.

Sobels: Can you use deficiencies in yeast and then demonstrate that within the region covered by the deficiency the meiotic effect disappears?

Magni: Yes, we have done such an experiment. A long deletion is available in *Saccharomyces cerevisiae* which covers locus *thr4*. Strains homozygous for the allele *thr4-1* show a 30-fold increase of reversion rate to threonine independence during meiosis. Diploid strains heterozygous for the deletion and consequently hemizygous for the mutant gene *thr4-1* do not show any variation of mutation rate between mitosis and meiosis. This shows that pairing in the region of the revertant locus is necessary for the meiotic effect.

Apirion: Can you tell us more about the "selfing" experiments Demerec did?

Magni: In *Salmonella* Demerec (1962) showed that the mutant *argA162* when transduced to a bacterium carrying the same allele reverts with higher frequency than in non-transduced control cells. He called the increase of reversion associated with DNA integration "selfing". He tested this phenomenon on 201 independent mutants and found that about 40 per cent showed the increase of reversion. His first interpretation was based on a typical unequal crossing-over model, but then he changed his mind. He found that deletions of the specific locus were not abolishing the increase of mutations in the transduced bacteria. He preferred as an explanation gene instability caused by the transducing phage (Demerec, 1963). The length of the deletions was not reported in his papers.

Apirion: We know for instance that in *E. coli* the *lac* region can be translocated by a recombinational event if there is a proper deletion in the chromosome. So the general problem that arises is what is the minimum size of recognition which is necessary for successful recombination? Selfing could also result from recombination after pairing of chromosomal regions which are only partially homologous.

Clarke: You said that nonsense suppressors show a meiotic effect. How do you explain this in terms of the structure and function of the tRNA molecule?

Magni: When I first presented these results (Magni and Puglisi, 1966) it was not yet proved that super-suppressors in yeast were nonsense suppressors and that they were mutations in tRNA. Our argument was the following: mutations from inactive to active suppressor (*su*→*Su*) show a meiotic effect and therefore they should be of the base insertion or deletion type. In addition super-suppressors are all dominant. These two facts rule out the possibility that the mutation (*su*→*Su*) affects a gene coding for a protein (activating enzyme, ribosomal protein, methylating or tRNA modifying enzyme), because a frame shift would knock out the synthesis of any protein and the mutation could not be dominant. The conclusion is therefore that mutations *su*→*Su* are occurring in a gene coding for an RNA and preferably tRNA. In reply to your specific question, an insertion or deletion could theoretically change the specificity of a tRNA as regards its capacity for recognizing a new codon in two ways: (1) the insertion or deletion occurs in the anticodon region, changing the structure of the anticodon triplet; (2) the recognition codon-anticodon is much more complex than we now believe, and a change in the nucleotide sequence at tRNA in a position other than the anticodon could affect the specificity of recognition.

Auerbach: Did you test whether any of those mutations were connected with recombination?

Magni: Unfortunately when we did this experiment none of the suppressors was mapped in *Saccharomyces cerevisiae*.

Auerbach: Did you not say that suppressors as a class show a broth effect? The cell in meiosis differs metabolically from the cell in mitosis, probably as much as cells grown on broth differ from cells grown on minimal medium. Might this not have something to do with your observations?

Magni: The only difference between mitosis and meiosis is the sporulation medium and this cannot be avoided. But we have shown that some mutants exhibit a meiotic effect and others do not. After the cell division during which the mutational event had occurred both mitotic and meiotic cells were seeded on the same medium.

Auerbach: This is not what I meant. You have shown very beautifully for the other types of mutation that the meiotic effect really is correlated with recombination, but one cannot from this conclude that every time one has a meiotic effect this must be due to recombination. It could also be due to the different metabolic environment.

Magni: Of course we can claim a relationship between mutation and recombination only in those cases where recombination was tested. On the other hand I doubt whether the external environment could account for our results.

Auerbach: I think it will be necessary to find out whether these suppressors arise from recombination.

Dawson: When one is comparing a replication cycle with a meiotic cycle, one is comparing unpaired chromosomes that are replicating with paired chromosomes that are replicating. In addition there is recombination in the meiotic cycle. Also there is no reason to assume that other physiological conditions of the cell are necessarily the same in the two cycles. So when one has a difference between mutation in a meiotic cycle and mutation in a mitotic cycle many possible differences could be the basis of this difference in mutation.

It might be interesting to look at the different mutation rates in the different *rec* mutants of *Neurospora*. There are two-element control systems which affect the frequency of recombination in particular small regions of the genome. In the different *rec*$^-$ and *rec*$^+$ mutants there could also be different mutation rates, on the basis of your arguments, in these particular regions.

Magni: The fact that the physiological conditions are different was the first point we analysed. Our experiments with deletions seem to me to reduce the probability that the physiological conditions are relevant, because in the experiment with the chromosomal deletion the biochemical conditions of meiosis were identical in the two strains with and without the deletion and nevertheless we observed a great difference in mutation rates. The only difference between the two strains was the lack of pairing and recombination in the one which did not show the meiotic effect. Therefore we are convinced that pairing and recombination are the major cause of the increase in mutation rate observed during meiosis.

Auerbach: You have proved your hypothesis fully for this class of mutation but not yet for the super-suppressors. It would be nice if you could do this.

Magni: It would be nice, yes. Theoretically it can be done now, as we know how to locate a good number of suppressors.

REFERENCES

DEMEREC, M. (1962). *Proc. natn. Acad. Sci. U.S.A.*, **48**, 1696–1704.

DEMEREC, M. (1963). *Genetics, Princeton*, **48**, 1519–1531.

MAGNI, G. E., and PUGLISI, P. P. (1966). *Cold Spring Harb. Symp. quant. Biol.*, **31**, 699–704.

MEGNET, R. (1966). *Experientia*, **22**, 151.

ZIMMERMANN, F. K., SCHWAIER, R., and LAER, U. VON (1966). *Mutation Res.*, **3**, 90–92.

GENERAL DISCUSSION

ELIMINATION OF UNEXCISED PYRIMIDINE DIMERS BY GENETIC RECOMBINATION

Devoret: The fate of a u.v.-damaged self-replicating unit—the F-*lac*$^+$ episome—transferred to various F$^-$ recipient cells deficient in recombination and/or pyrimidine dimer excision has been investigated in *Escherichia coli* K12 (Devoret *et al.*, 1969). The aim was to determine by genetic means the possible structure of the transmitted DNA. One of the conclusions to be derived from this work is that the deleterious effects of unexcised pyrimidine dimers—which prevent the expression of the transferred *lac*$^+$ genes—can be circumvented by genetic recombination in the recipient cell. Marker rescue of the u.v.-damaged *lac*$^+$ genes takes place if the F$^-$ recipients are recombination-proficient Rec$^+$ and carry a piece of chromosome homologous to the transmitted operon.

Let me now report the first experiment in which the two following crosses were performed:

(1) U.v.-irradiated *uvrB*501 *recA*13 / F-*lac*$^+$ × F$^-$*rec*$^+$ *lacY*1

(2) U.v.-irradiated *uvrB*501 *recA*13 / F-*lac*$^+$ × F$^-$*recA*13 *lacY*1

The F-*lac*$^+$ donor cells—recombination (Rec$^-$) and excision-repair deficient (Uvr$^-$)—were u.v.-irradiated and then mated with Rec$^+$ and Rec$^-$ recipient (F$^-$) cells. Both Lac$^-$ females carried *lacY*1, a very small deletion in the lactose-permease gene. The yield of the Lac$^+$ sexduced colonies (as seen in Fig. 1) is drastically reduced in the Rec$^-$ recipients with increasing u.v. doses. Under the experimental conditions used any possible killing of the recipient cells during mating was eliminated.

To examine whether the Lac$^+$ colonies from Rec$^+$ or Rec$^-$ recipients still carried a functional F-*lac*$^+$ episome, the colonies were tested for sensitivity to male-specific phage MS2 as well as for their ability to act as secondary F-*lac*$^+$ donor cells. At high u.v. doses, all the Lac$^+$Rec$^-$ colonies carry a functional F-*lac*$^+$ episome whereas all the Lac$^+$Rec$^+$ colonies behave as females. The latter colonies should therefore be recombinants.

To prove this point further we performed two crosses in which the same u.v.-irradiated *uvrB*501 *recA*13 / F-*lac*$^+$ donor cell was mated with two Rec$^+$ recipients differing in the structure of the chromosome at the *lac* operon. The *lac* region was present on the chromosome of the first recipient which carried a β-galactosidase *lacZ* point mutation. This recipient had been derived by transducing the second strain which had a deletion of the entire *lac* operon (Δ*lac*). The yield and the female phenotype of the Lac+ colonies formed at high u.v. doses in the *lacZ* recipients are identical with those obtained from the *lacY1* recipients. In contrast, the yield of Lac+ colonies in Δ*lac* recipient cells tends to parallel that of the Lac+Rec− colonies formed in cross (2). Moreover, most of the colonies carry a functional F-*lac*+ episome.

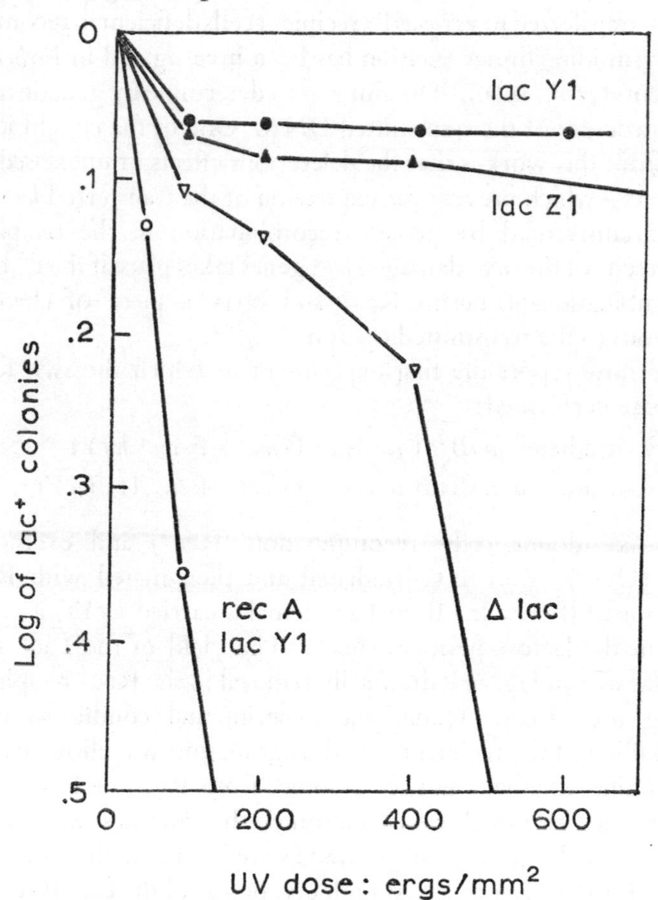

FIG. 1. (Devoret). Ordinate (log scale) shows the relative numbers of Lac+ colonies formed in various non-irradiated F− recipient bacteria (○: *recA lacY1*; ●: *recA+ lacY1*; ▽: *recA+* Δ*lac*; ▲: *recA+ lacZ1*) after mating for 30 min at 37° with *uvrB501 recA13*/F-*lac*+ donor cells previously exposed to increasing u.v. doses (abscissa). The ratio of male to female cells in the mating mixture was about 1:10 to avoid multiple *F-lac*+ transmission.

Thus, the Lac^+F^- colonies in cross (1) arise from recombination between the damaged episome and the recipient chromosome.

As to the question of whether pyrimidine dimers are repaired when transferred to the recipient cells, we had previously shown (Devoret and Rörsch, cited by Mattern, van Winden and Rörsch, 1965; Pardee and Prestidge, 1967) that the recovery of Lac^+ zygote colonies from a u.v.-irradiated F-*lac*$^+$ episome transferred to F^- recipients was not at all influenced by the Uvr phenotype of the female cells. This was confirmed when crosses similar to (1) and (2) were performed with females carrying an additional deficiency in excision repair. The results were identical to those obtained in crosses (1) and (2) (see Fig. 1). To account for all of these facts, we suggest that the structure of the transferred u.v.-damaged F-*lac*$^+$ episome is such that it is not susceptible to pyrimidine-dimer excision.

In conclusion, marker rescue of a u.v.-damaged gene can occur in the absence of pyrimidine-dimer excision through a process of genetic recombination.

Auerbach: So this is repair of a potential mutation which otherwise would be lost?

Devoret: It is inactivation of the *lac*$^+$ operon. When the data (LD_{37} is about 13 erg/mm^2) obtained from cross (2) were considered in the light of the findings of Rupp and Howard-Flanders (1968) concerning the mean molecular weight of the DNA stretched between pyrimidine dimers, and those of Freifelder (1968) on the size of the F-*lac*$^+$ episome, we were able to estimate that the F-*lac*$^+$ episome transferred to *recA* recipients is inactivated by about one pyrimidine dimer.

Kimball: There must be a fair number of dimers originally in the *lac*$^+$ gene in the male; nevertheless it can function adequately in the female. That seems to be the conclusion.

Bridges: Did you use an Hcr^- strain for the female in these experiments?

Devoret: Yes. We used recipients carrying all the various combinations of Rec and Uvr deficiency.

Clarke: Have you used any *fil* or *lon* mutants?

Devoret: We used typically excision- and recombination-deficient mutants.

Auerbach: How do you think the dimers were resolved by recombination?

Devoret: We can imagine that the u.v.-irradiated *lac* genes recombine with the recipient chromosome, either by a single cross-over if the F-*lac*$^+$ episome becomes circular or by a double cross-over if it stays as a linear piece. But let me recall a more general hypothesis put forward by Rupp and Howard-Flanders (1968) as to the question of how dimers can be resolved by recombination. The hypothesis is based upon the facts that:

(*a*) The 37 per cent survival dose in excision-deficient bacteria corresponds to the presence of about 50 dimers per chromosome whereas in excision- *and* recombination-deficient bacteria it is about 1·5 dimers per genome.

(*b*) DNA replication goes on in excision-defective bacteria after u.v. irradiation.

(*c*) Discontinuities are formed in the newly synthesized daughter strand; they seem very likely to be opposite dimers.

If bacteria deficient in excision repair can survive 50 dimers per chromosome but cannot do so if they are also recombination deficient, there should be a mechanism that replaces the daughter-strand defect with intact DNA through a recombinational process.

Figure 9b in Rupp and Howard-Flanders' paper is as below:

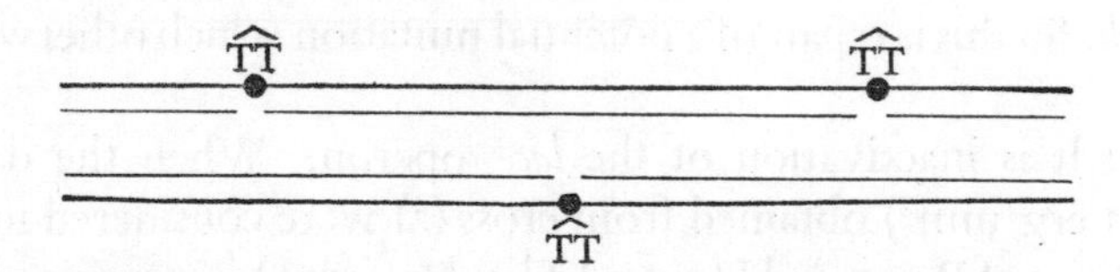

It is apparent that the region containing a dimer in one duplex is intact in the sister duplex. Recombination between two sister duplexes would permit the recovery of the potentially lost information.

A post-replication repair mechanism of this type could complement the excision repair process in wild-type cells. While excision repair would be effective before replication, the post-replication genetic repair mechanism would act on abnormalities in the daughter strands. Thus if the replication apparatus passes an unexcised dimer, defective or missing information in the daughter strand might subsequently be replaced with genetic information derived from the sister duplex. Complete genetic information could be restored by a recombination process.

Auerbach: In both strands or in one?

Devoret: In both strands. Only one strand per dimer is lost so it is a kind of conversion, if you like.

Maaløe: When you deal with a single dimer, why do you want to postulate a replication fork? The endpoint of your assay is survival or not survival, and if you place the dimer after the fork there is little need for recombination.

Rörsch: The dimers are still allowed to pass over the fork.

Maaløe: I am just saying that when the thymine dimer is in one arm in the portion that has been replicated survival could be assured by the undamaged arm.

Clarke: The recombinational repair needs two dimers and two gaps.

Devoret: I wanted to simplify this.

Loveless: I don't get the argument about whether the thing is in the part which has already replicated. The problem is only going to come up in the next cell generation.

Rörsch: All you need is to restore one intact genome. This could also be constructed by recombination between the two nuclei which are always present in each cell.

Clarke: Then we should never expect a strictly exponential killing curve with u.v.

Devoret: Yes, because you make breaks.

Maaløe: Dr. Rörsch could be right, but we now have evidence that when one round of replication is finished in an *E. coli* cell with a single nucleus, physiological division actually takes place. Thus, during the remaining part of the division cycle the two genomes are carried in separate compartments, so that recombination might never be permissible between sister nuclei.

Kaplan: I understand that recombination has to happen because the thymine dimer destroys the genetic information in this strand. Is it also possible to assume that there is an enzyme in the cell which splits the thymine dimers and thus restores the genetic information (it would be a special enzyme, working only opposite a gap, of course)? It would not then be necessary to assume recombination: it would be a repair process.

Devoret: Do you mean that if you use photoreactivation, for instance, you restore a function which has been lost?

Kaplan: Not photoreactivation, because there is a gap, and as far as I know the photoreactivating enzyme does not work in single-strand DNA.

Devoret: This experiment has been done in the F-*lac*$^+$ system I described. Photoreactivation of recipient cells which have received a u.v.-damaged F-*lac*$^+$ episome enhances the yield of recovered Lac$^+$ colonies, as demonstrated by R. S. Cole and P. Howard-Flanders (personal communication).

Kaplan: Of course when you split the dimers by photoreactivation the wild type is not restored. But when you have a gap opposite the thymine dimer the genetic information is destroyed in this daughter molecule.

Devoret: It is unavailable in one daughter molecule but it is available in the second one.

Kaplan: Yes, but to make it available in this strand (without recombining out of the dimer) an enzyme could split the dimer and so restore the original wild-type information.

Devoret: We are talking about a bacterium which can't excise. Pyrimidine dimer excision deficiency was determined by biochemical means and not by host-cell reactivation alone.

Bridges: Could we move on from this to mutagenesis? This model of Rupp and Howard-Flanders (1968) was very interesting, and it didn't take very much imagination to suggest that it is in the filling up of such gaps opposite dimers that mutations could occur. Evelyn Witkin, Munson and myself suggested this a year or so ago (Witkin, 1967; Bridges, Dennis and Munson, 1967). What in fact is the evidence in support of this? The first thing—and this is relevant to what Professor Kaplan was saying—is that the dimers persist as dimers. The evidence is not entirely biochemical because one can only do the biochemistry at quite high dose levels. We have shown that the photoreversibility of mutations can persist for three or more generation times after u.v. (Bridges and Munson, 1968). The dimer is still there as a photoreversible entity going through DNA replication cycles, and we have shown that if DNA replication is stopped then the mutations stop appearing. So it seems to me that it is the process of replication which is producing mutations. What happens as these dimers go through the replication complex? The frequency with which mutations are induced in an ochre triplet that we work with is sufficiently low for one to conclude that the probability of a mutation appearing when a pyrimidine dimer at that triplet goes through the replication point is of the order of 1 to 3 or 5 per cent. This is so low that it cannot be due to a random insertion of bases. The information to fill the gap in the strand opposite the pyrimidine dimer has to come from somewhere and the only place it can come from is the other daughter duplex. That is consistent with a recombinational repair model.

We determine when a mutation is produced by the loss of its photoreversibility. Once it has changed from being a pyrimidine dimer to being something else, from that point onwards all the progeny from that duplex are mutant. Again this is consistent with the Rupp and Howard-Flanders model. Is recombination involved? It is hard to see what else could be involved but we have further information about the *exr* gene in *E. coli*. Witkin (1967) has shown that Exr^- strains are not mutable to a significant extent by u.v. They are also slightly more sensitive to u.v. and ionizing radiation. Moreover she has found that Exr^- bacteria are also Rec^- to a certain extent. Here we have a correlation between recombination ability,

ability to mutate, and ability to cope with dimers which persist through replication.

Apirion: Is this true also for a Uvr$^-$ strain?

Bridges: Yes.

Apirion: The phenomenon should be called avoidance by recombination, rather than excision by recombination. Does it mean that the thymine dimer persists forever, so to speak, in such strains?

Bridges: It doesn't seem to persist forever. We can explain our curves on the assumption that nine or ten dimers per generation time or replication cycle disappear or lose their photoreversibility. This is a very low rate, and we have no idea whether this is due to a very small amount of excision, or to a very small amount of splitting. I tend to favour the latter because this bacterium has a photoreactivating enzyme, and there might well be sufficient quanta of energy within the cell to enable a few splittings to occur, even in laboratory darkness.

Clarke: You had a system called low temperature mutation loss, active on u.v. and X-ray damage. Do you think this may be recombination repair?

Bridges: It is possible to explain it in terms of recombination. We get loss of mutations at low incubation temperatures after both u.v. and X-rays, and we have also shown that the Exr mutation affects both u.v.- and X-ray-induced mutation. One may therefore hypothesize that low temperature mutation loss is a reflection of the probability of the EXR system making a mistake at different temperatures.

Kubitschek: I think that this can be explained by a master-strand model.

Clarke: There seems to be a very great need for synchronous culture studies, to see whether some mutagens at least act preferentially before, at or after the replication fork. I would predict that X-rays would only produce mutations after the replication fork. There is some evidence already that nitrosoguanidine goes for the replication fork, preferentially.

Kubitschek: Ann Finney, R. Krisch and I used synchronous cultures to see whether ultraviolet light kills cells more efficiently before or after replication of the DNA. This was done in very slowly growing chemostat cultures, where there was a long G1 period before the synthesis of DNA. In *E. coli* THU and in a B_{s-1} strain there was no significant difference, at the 6 per cent level, in the survival curves at various times during the cycle.

Kimball: One can certainly induce mutations, not necessarily in *E. coli* but in some systems, in G1; that is, before the replication fork has passed.

Maaløe: Why do you believe that X-rays selectively damage the already replicated portion of a genome, Dr. Clarke?

Clarke: Because I don't like the idea of recombination repair between separate genomes.

Böhme: The association of recombination with the production of "spontaneous" mutations has been shown repeatedly (see Magni, this volume, p. 186). Evidence is now accumulating which characterizes defects in repair mechanisms as a further factor in the origin of spontaneous mutations.

Possible connexions between the function of cellular repair mechanisms and frequency of spontaneous mutability have repeatedly been discussed (e.g. Hanawalt and Haynes, 1965; Zamenhof, Heldenmuth and Zamenhof, 1966). Mutants have now been described in both prokaryotes and eukaryotes in which a defect of a repair mechanism seems to be coupled with a change in the frequency of spontaneous mutability (Table I). Among these are *rec*⁻ mutants as well as types with normal behaviour in recombination.

TABLE I (BÖHME)

MUTATOR EFFECTS ASSOCIATED WITH REPAIR DEFECTS IN MICRO-ORGANISMS

Organism	*No. of mutants*	*Reported phenotype**	*Reference*
B. subtilis	1	Rec	Prosorov and Barabančikov, 1967
P. mirabilis	7	Uvr	Böhme, 1967, 1968
E. coli K12	2	Uvr	Mohn, 1968
E. coli B	1	Rec	Kondo, 1968
N. meningitidis	1	Rec	Jyssum, 1968
S. cerevisiac	2	Uvr	Zakharov, Kozhina and Fedorova, 1968
S. cerevisiae	2	Uvr; Exr	von Borstel, 1968

*Classification according to Rörsch *et al.*, 1967.

Recently Hill (1968) has pointed out that up to now in none of these cases is there unequivocal evidence that both mutator effect and repair defect—e.g. increased u.v. sensitivity—are pleiotropic effects of one and the same mutation and not due to two independent mutational events. In the Mut mutants of *P. mirabilis* (Böhme, 1968), two arguments support the assumption that both mutator effect and repair defect are due to a single mutation. Firstly, from 27 u.v.-sensitive mutants seven Mut mutants have been isolated which show a completely identical range of characters (Table II). If mutator effect and repair defect were due to independent mutations, one would expect that the mutator trait might also be found in combination with other repair-defective mutant types. Table II shows that the mutator trait can only be seen in one type of repair mutant; in other words, there is

TABLE II (BÖHME)

PHENOTYPIC CHARACTERIZATION OF U.V.-SENSITIVE MUTANTS OF *Proteus mirabilis*

Strain	*Uv*	*HN2*	*Hcr*	*Coff*	*Mut*	*Phde*	*X-R*	*Ems*	*Rec*	
273	+	+	+	+	+	+	+	+	+	Parent strains for mutant selection
758	+	+	+	+	+	+	+	+	+	
671	−	−	+	+	+	+	+	+	+	Uvr
674	−	−	+	+	+	+	+	+	+	
673 677	−	−	−	−	+	+	+	+	+	Hcr
678 679	−	−	−	−	+	+	+	+	+	
683 684	−	−	−	−	+	+	+	+	+	
685 686	−	−	−	−	+	+	+	+	+	
689 690	−	−	−	−	+	+	+	+	+	
692 694	−	−	−	−	+	+	+	+	+	
696 697	−	−	−	−	+	+	+	+	+	
667	−	−	+	−	−	+	+	+	+	Mut
668	−	−	+	−	−	+	+	+	+	
669	−	−	+	−	−	+	+	+	+	
670	−	−	+	−	−	+	+	+	+	
680	−	−	+	−	−	+	+	+	+	
688	−	−	+	−	−	+	+	+	+	
695	−	−	+	−	−	+	+	+	+	
687	−	−	+	+	+	+	−	−	+	
675	−	−	+	+	+	−	−	−	+	Exr
693	−	−	+	+	+	−	−	−	+	
672	−	−	+	+	+	−	−	−	−	Rec

Abbreviations:
Uv^- = sensitivity to u.v. light
$HN2^-$ = sensitivity to di-(2-chloroethyl)methylamine
Hcr^- = decreased host cell reactivation
$Coff^-$ = resistance to caffeine post-treatment after u.v.
Mut^- = increased spontaneous mutability
$Phde^-$ = sensitivity to photodynamic inactivation
$X\text{-}R^-$ = sensitivity to X-rays
Ems^- = sensitivity to ethyl methanesulphonate
Rec^- = decreased recombination capacity

no reason to expect an increased mutability for any repair-defective mutant.

Secondly, for more direct evidence we isolated u.v.-resistant revertants from Mut strains and examined their mutability. Five u.v.-resistant revertants have been isolated from Mut strain PG 670. None of them differed from the wild type with respect to u.v. resistance. All showed a drastically reduced mutant frequency which differed only slightly from the mutant frequency of the wild type. Results of a typical experiment are given in Table III.

TABLE III (BÖHME)

FREQUENCY OF SPONTANEOUS THREONINE (*thr*) PROTOTROPHIC AND STREPTOMYCIN (*str*)-RESISTANT MUTANTS (MUTANTS/10^8 CELLS) OF U.V.-RESISTANT REVERTANTS OF MUT STRAIN PG 670

Strain		$thr^- \rightarrow thr^+$	*str-s→str-r*
PG 273	Wild type*	0·03	0·12
PG 670	Mut	10·02	4·82
PG 670/1	U.v.-resistant revertants of PG 670	0·077	0·12
PG 670/2		0·030	0·029
PG 670/3		0·14	0·054
PG 670 4		0·13	0·11
PG 670/5		0·058	0·13
PG 670/6		0·067	0·12

*With respect to u.v.-resistant markers.

Of greater interest is the more precise characterization of the mutator effectiveness as such. The first question we tried to answer concerns the relative amount of mutator effectiveness. A comparison between "spontaneous" mutant frequency in one of the Mut strains (PG 688) and the mutant frequency induced by ethyl methanesulphonate and hydroxylamine in the u.v.-resistant parent strain showed that the mutator effect corresponds roughly to the effectiveness of a 20-minute treatment with EMS (survival: 45 per cent) or a 30-minute treatment with hydroxylamine (survival: 2 per cent).

The second question raised concerns the specificity of the mutator effect. We were especially interested in finding out whether mutator-induced mutations can be reversed again by the mutator. Starting with one of the Mut strains (PG 688; thr^-; arg^-), we isolated strains with a third auxotrophy induced by the mutator and determined their reversion rate. All four strains studied also showed a drastically increased reversion rate for the mutator-induced auxotrophy which appears to be even higher than the mutator-induced reversion rate of those alleles not induced by the mutator. These preliminary results indicate that the mutator effect is unspecific, as opposed to the unidirectional action of the mutator in the Treffers strain of *E. coli* (Yanofsky, Cox and Horn, 1966). At the moment we can only speculate about the mechanism of the mutator effect. However, we have some reason to assume that the mutations in Mut strains are induced by an incorrectly proceeding repair process rather than representing the consequence of a missing repair of pre-existing lesions. (The experiments summarized here were done in cooperation with Dr. B. Adler and Mr. W. Witte.)

Magni: In quantitative analysis of any kind of mutation, either spontaneous or induced, it is very important to know the exact genetic background of the organism used. The presence of a mutator gene could easily affect any quantitative estimate, especially when the mutator is specific at the molecular level, such as the mutator described by Treffers, Spinelli and Belser (1954) that induced only transversions (Yanofsky, Cox and Horn, 1966) or the mutator of von Borstel (1968) which specifically induces frame shifts.

Böhme: This argument, important as it may be for your organism, is without any relation to our material; our strains are isogenic and differ only in the mutation selected for.

Magni: You are right. But in higher organisms such as yeast and *Neurospora* strains are continuously crossed with each other and one must be very careful not to introduce mutators in the cross.

Kaplan: The mutator strain in *E. coli* K12 which was discovered by Dr. Mohn and myself (Mohn and Kaplan, 1967; Mohn, 1968) seems similar to the one described by Dr. Böhme. But one difference is that our mutator is very specific for the gene which is mutated. This is a mutation of *met*$^-$ to *met*$^+$. We also tried resistances to different phages and different antibiotics which certainly consist of many different types of forward mutations. They are not influenced in their mutability; thus the mutator seems rather specific for certain types of spontaneous mutations.

Dawson: Has this particular mutator gene been plotted?

Kaplan: We know that it is not close to the gene whose mutations are being induced.

Dawson: So this might be a two-element control system for mutations. This would mean that there is a recognition region in the gene in which the mutations are being induced and the mutator gene, elsewhere in the genome, is a regulator gene.

Kaplan: It could be something like that.

Rörsch: In 1965 we described four different *uvr* mutants, with mutations in the genes *uvrA*, *uvrB*, *uvrC* and *uvrD* (formerly designated *dar*-3, *dar*-1, *dar*-5 and *dar*-2, respectively: van de Putte *et al.*, 1965). The *uvrD* mutants differ from the others in being u.v. sensitive, which is not accompanied by a strongly reduced host cell reactivation. Having learned about Mohn and Kaplan's mutator strain (personal communication) we tested the frequency of spontaneous reversions of auxotrophic markers and found that the *uvrD* mutants are highly unstable in this respect. Recently Ogawa, Shimada and Tomizawa (1968) described a *uvrD* type mutant and established that it shows an impaired repair replication. This can account for the high mutability of such strains.

Dawson: Does this only affect one particular gene?

Rörsch: It is not specific; several auxotrophic mutations are affected in our *uvrD* mutants.

Kaplan: Our mutator is very specific. Many forward mutations are not affected; *met*$^-$ to *met*$^+$ is the only one we know of.

Grossman: If it is assumed that the mutated gene controls an aberrant polymerase and experiences modified recognition, then if such mutants are treated with agents such as hydroxylamine, one should observe a different spectrum of activity by these agents.

Kaplan: No. This mutator has no influence on the mutability induced by hydroxylamine, ethyl methanesulphonate, nitrosoguanidine, 5-bromouracil, and so on.

Magni: May I propose a very naive explanation for your data, Professor Kaplan? Let us assume that the specific mutation which is affected by your mutator is a frame shift. In other words your mutator is a mutator gene of the same type as von Borstel (1968) has found in yeast, i.e. a specific mutator for frame shifts. If most forward mutations are base substitutions, as is very probable, they will not be affected by your mutator.

Dawson: When you said this mutator affected a particular gene, Professor Kaplan, is it the reversion of a particular allele of that gene or does it generally induce mutation in that gene?

Kaplan: It is just this special allele *met*1$^-$ which was in the strain already when the mutator was introduced.

Böhme: The explanation which assumes a mutant polymerase is difficult to apply to your case, since your mutator strain is *hcr*$^-$. In our Mut mutants the capacity for host cell reactivation is unreduced.

Kaplan: The *hcr*$^-$ capacity was tested with phages T1 and and λ. T1 has a decreased survival after u.v. in the mutator strain, λ an increased prophage induction. That is the difference between these two systems.

Rörsch: Are you sure that you are dealing with a single mutation or with two mutations. Did you test whether the *hcr*$^-$ property could be separated from the property that affects the mutability?

Kaplan: It is not yet clear. But the mutations must be closely linked, for in conjugation they go together.

Pollock: We have recently found an interesting unstable mutation in *B. cereus* in which every cell of the population can undergo a certain mutational event in a matter of an hour or two (Pollock and Fleming, 1968). This event involves a complete "switch-on" of the production of the enzyme penicillinase. Many of the features of this instability seem to be very similar to ones previously studied in other organisms with more respectable genetics, such as the well-known phase variations and those Professor Dawson and Dr. Smith-Keary (1963) have been studying with respect to proline auxotrophy in *Salmonella*, as well as certain mutational events studied by Barbara McClintock (1961) in maize, concerned mainly with anthocyanin production.

This unstable strain is really a double mutant. The original wild type is inducible with respect to penicillinase, and undergoes a constitutive mutation to complete derepression of the penicillinase gene. Now this spontaneous magno-constitutive mutant (strain 569/H) further mutates spontaneously to partial or complete repression of its penicillinase gene. Amongst the many total loss mutations that occur, there is one particular unstable type which turns up regularly though relatively infrequently. This mutant type (referred to as H/14) reverts back to the fully derepressed penicillinase-positive state of its parent strain 569/H at a very high rate, quite spontaneously. Even under ordinary growth conditions this mutant is quite clearly highly unstable. It starts completely penicillinase-negative and highly penicillin-sensitive, switching back to a full expression of the gene, with complete recovery of penicillin resistance.

We discovered that this reverse mutation could be induced, so that *all* the cells in a population were affected, by means of two non-specific types of environmental stimulus. The first was raising the temperature from 35°–37°C (the normal growth temperature) to between 42°–45° (at the upper limit of which cells hardly grow at all). The other was the addition of chloramphenicol (20 μg/ml). In both these conditions there may be no increase, nor indeed any decrease, in viable count. I have called this phenomenon a mass genetic conversion. If grown at 25° nothing apparently happens: at 35° the culture reverts slowly (i.e. in 24 hours) to the derepressed state. At 44°, however, after an absolute lag phase of about 90 minutes, there is a rapid switch-over, lasting only about 60 minutes, of all the cells to the completely derepressed state. These conversion kinetics are almost identical when the inducing stimulus is chloramphenicol.

The converted population shows considerable, though not complete, genetic stability. The switch, though overwhelmingly polarized towards the derepressed state, is reversible. The convertants are phenotypically indistinguishable from the original parent strain 569/H, and will throw off loss mutants (including occasionally the unstable, H/14, type) in just the same way, so that populations can, in a sense, oscillate between the fully repressed and fully derepressed states, perhaps indefinitely. The underlying event cannot therefore be simply the loss of a plasmid or any type of deletion of exclusive genetic information.

As far as can be judged, since enzyme production at 44° can be measured quite satisfactorily, there is no phenotypic lag in the expression of this mutational event. In other words, penicillinase can be detected after 90 to 100 minutes and increases rapidly at a more or less exponential rate. Not more than an hour, perhaps less, intervenes between the event by which cells become genetically committed to the derepressed state and the physiological switch-on of penicillinase production.

A possibly relevant observation is that this conversion, whether brought about by chloramphenicol or by raised temperature, is up to 70 per cent inhibited by nalidixic acid (10 μg/ml). This substance is supposed to be a specific inhibitor of DNA synthesis (Barbour, 1967), which may, therefore, play some part in the process.

Whatever may be the nature of the event which underlies this conversion, there are two points worth bearing in mind. One is that the switch involved is purely regulatory: it is a switch which determines whether or not the gene is to be fully expressed. Secondly, the fact that the switch is turned on by addition of chloramphenicol *or* by raising the temperature suggests that a protein inhibits the conversion event in the normal strain and that in the unstable mutant this protein may be abnormally temperature-sensitive. It is difficult to interpret the similar action of chloramphenicol and raised temperature without bringing in a thermolabile protein somewhere or other. Indeed, on that hypothesis one could predict that if, in fact, there were an unstable repressor of conversion, raising the temperature would destroy this repressor whereas adding chloramphenicol would inhibit further production. So that if the two were combined, if this hypothesis is correct one would expect greatly to decrease the conversion lag, which could be interpreted as being the time necessary to allow the repressor to fall below a threshold level. Indeed, treatment of an H/14 culture with chloramphenicol at 45° reduced the conversion lag markedly, from about 90 minutes to less than 35 minutes.

Thus, the thermolabile protein repressor hypothesis appears not unreasonable though it tells us little about the underlying conversion event.

One possible interpretation might be that what is involved is a recombinational translocation, requiring DNA synthesis, which effects a reversible approximation/separation of the penicillinase structural gene to/from a *cis*-acting promoter-type region necessary for its proper expression.

This is of course highly speculative and until we can undertake some proper genetic studies (as would at last seem to be feasible, thanks to recent work by Thorne, 1968) we can't even show that DNA is involved. Professor Dawson and others who have worked on highly unstable genetic states in other organisms may be able to offer some suggestions, or perhaps see where this particular piece of evidence might help towards understanding unstable mutations in organisms which have more respectable genetics.

Dawson: It is extremely difficult to get very far, as you say, until one knows more about the genetics. One doesn't know whether this change is at the gene itself or at a regulator gene. McClintock (1965) has described in maize a situation where a regulator operates so that a particular gene is switched on synchronously at a particular time of development in numerous cell lines of the endosperm. Formally one could think of your situation along similar lines. The distinction here is the very high levels of instability, which can switch mutation on and off.

Maaløe: I still believe that the phenomenon you observed has to do with the going on and coming off of an episome.

Pollock: If there is something essential on an episome, whether it is a repressor of a structural gene or a promoter for it, or the structural gene itself, one would expect a high rate of irreversible loss, either of the repressor or the promoter or structural gene—whichever was on the episome. But we don't find that. There is no indication that any genetic factor specifically affecting penicillinase production in this strain of *B. cereus* is easily and irreversibly lost. Where *relatively* frequent loss mutations occur (i.e. from the magno-constitutive [569/H] to the negative or micro-constitutive state, where there is practically no enzyme at all) they are always reversible.

Devoret: Schwartz (1965) has found a *lac*⁻ mutation in *E. coli* K12 which has almost the same features as the one you describe.

Maaløe: To me the simplest way of coordinating your suggestion that a control system is switched on or off with the existence of a semi-stable genetic state would be to say that when the element X, which may or may not carry the penicillinase cistron, is inserted into the chromosome, synthesis of the enzyme comes under the control of a different regulator system.

Clarke: One naive explanation for your results is that first you have the operator with the repressor on it, and the transcriptase can't act; the repressor comes off or is destroyed, the transcriptase then acts and goes at such a rate that the repressor cannot reattach.

Pollock: Formally, that is the same as saying that the repressor is necessary for its own synthesis. That possibility cannot be excluded until it is shown that the derepressed genetic state is transmissible by mechanisms (transformation, transduction or conjugation) involving transfer only of DNA.

Kimball: There are formal similarities with the antigen systems in *Paramecium*. In these one can turn on and turn off a system somehow, in a semi-permanent way. One can grow two different antigenic types for many cell generations side by side under the same conditions and they remain distinct. Yet either can be transformed into the other in appropriate conditions. There is a formal similarity in the system that probably doesn't involve a mutation in the ordinary sense of that word.

Dawson: Except that in the strain described by Pollock the apparent mutation rate is high in one direction only.

Pollock: I don't quite see the point in arguing about a recombinational event involving episomes. This would just be a special case of recombinational events in general. It is possibly true; but in fact there is no evidence that either of the elements possibly implicated is on an episome. I think it is premature to consider a special case. What we would like to know is whether a recombinational event is involved *at all*. We even need to know, as Dr. Kimball has challenged, whether it really involves DNA or is due to some autocatalytic metabolic cycle as Dr. Clarke suggests. We have got to get an answer to this. The only thing that can be said so far is that nalidixic acid is a specific inhibitor of induced conversion; this suggests that perhaps DNA synthesis is involved and that, in so far as DNA synthesis is necessary for recombination, the latter process may possibly be implicated.

Apirion: The episome idea is quite adequate here, whether it is true or not. The best argument against it is why is there no segregation? This could be explained if the same episome carries at least one essential gene for the cell.

Pollock: It just seems to me to be unnecessary to argue at the moment whether it is due to an episome or not. It could perfectly well be an intrachromosomal recombination, and I think we want first to know whether it is *either* of these.

Apirion: It is difficult to envisage a mechanism that could account for 100 per cent induced recombination in a specific chromosomal region.

Dawson: One would like to know how the system works in a *rec*⁻ strain.

Pollock: Of course, but we have first to get working in *B. cereus* a genetic system by which *rec*⁻ mutants can be recognized.

REFERENCES

BARBOUR, S. D. (1967). *J. molec. Biol.*, **28,** 373.
BÖHME, H. (1967). *Biochem. biophys. Res. Commun.*, **28,** 191–196.
BÖHME, H. (1968). *XII Int. Congr. Genet.*, Tokyo, **1,** 76.
BORSTEL, R. C. VON (1968). *XII Int. Congr. Genet.*, Tokyo, **2,** 124.
BRIDGES, B. A., DENNIS, R. E., and MUNSON, R. J. (1967). *Genetics, Princeton*, **57,** 897–908.
BRIDGES, B. A., and MUNSON, R. J. (1968). *Proc. R. Soc. B*, **171,** 213–226.
DAWSON, G. W. P., and SMITH-KEARY, P. F. (1963). *Heredity, Lond.*, **18,** 1.
DEVORET, R., GEORGE, J., THOMPSON, M., and HOWARD-FLANDERS, P. (1969). In preparation.
DEVORET, R., and RÖRSCH, A., cited by MATTERN, I. E., WINDEN, M. P. VAN, and RÖRSCH, A. (1965). *Mutation Res.*, **2,** 111–131.
FREIFELDER, D. (1968). *J. molec. Biol.*, **35,** 95–102.
HANAWALT, P. C., and HAYNES, R. H. (1965). *Biochem. biophys. Res. Commun.*, **19,** 462–467.
HILL, R. F. (1968). *Mutation Res.*, **6,** 472–475.
JYSSUM, K. (1968). *J. Bact.*, **96,** 165–172.
KONDO, S. (1968). *XII Int. Congr. Genet.*, Tokyo, **2,** 126–127.
MCCLINTOCK, B. (1961). *Am. Nat.*, **95,** 265.
MCCLINTOCK, B. (1965). *Brookhaven. Symp. Biol.*, **18,** 162–182.
MOHN, G. (1968). *Molec. gen. Genet.*, **101,** 43–50.
MOHN, G., and KAPLAN, R. W. (1967). *Molec. gen. Genet.*, **99,** 191–202.
OGAWA, H., SHIMADA, K., and TOMIZAWA, J. (1968). *Molec. gen. Genet.*, **101,** 227.
PARDEE, A. B., and PRESTIDGE, L. S. (1967). *J. Bact.*, **93,** 1210–1219.
POLLOCK, M. A., and FLEMING, J. (1968). *Microb. Genet. Bull.*, **29,** 12.
PROSOROV, A. A., and BARABANČIKOV, B. I. (1967). *Dokl. Akad. Nauk SSSR*, **176,** 1422–1424.
PUTTE, P. VAN DE, SLUIS, C. A. VAN, DILLEWIJN, J. VAN, and RÖRSCH, A. (1965). *Mutation Res.*, **2,** 97.
RÖRSCH, A., PUTTE, P. VAN DE, MATTERN, I. E., and ZWENK, H. (1967). *Radiat. Res.*, **19,** 771–789.
RUPP, W. D., and HOWARD-FLANDERS, P. (1968). *J. molec. Biol.*, **31,** 291–304.
SCHWARTZ, N. M. (1965). *J. Bact.*, **89,** 712–717.
THORNE, C. B. (1968). *J. Virol.*, **2,** 657–685.
TREFFERS, H. P., SPINELLI, V., and BELSER, N.O. (1954). *Proc. natn. Acad. Sci. U.S.A.*, **40,** 1064–1071.
WITKIN, E. M. (1967). *Brookhaven Symp. Biol.*, **18,** 17–55.
YANOFSKY, C., COX, E. C., and HORN, V. (1966). *Proc. natn. Acad. Sci. U.S.A.*, **55,** 274–281.
ZAKHAROV, I. A., KOZHINA, T. N., and FEDOROVA, I. V. (1968). *Dokl. Akad. Nauk SSSR*, **181,** 470–472
ZAMENHOF, S., HELDENMUTH, L. H., and ZAMENHOF, P. J. (1966). *Proc. natn. Acad. Sci. U.S.A.*, **55,** 50–58.

OBSERVED MUTATION FREQUENCY IN MICE AND THE CHAIN OF PROCESSES AFFECTING IT*

W. L. Russell

Biology Division, Oak Ridge National Laboratory, Oak Ridge, Tennessee

Until a few years ago it was commonly believed that radiation-induced gene mutation rate is linearly related to radiation dose, is independent of dose rate and dose fractionation, is not affected by the interval between irradiation of a particular germ cell stage and fertilization, and, in general, shows no evidence of repair of radiation-induced gene mutational damage. These conclusions, which were widely accepted as basic principles of radiation genetics, and were regarded, as late as 1958 (United Nations Scientific Committee on the Effects of Atomic Radiation) as applicable to man, were based largely on work with *Drosophila* spermatozoa. Work with mice over the past two decades has shown that the observed mutation frequencies after irradiation of spermatogonial or oocyte stages show fundamental differences from the *Drosophila* results. For example, there are marked effects of dose rate and dose fractionation. Furthermore, the responses of mouse spermatogonia, oocytes, and spermatozoa proved to be strikingly different in many more ways than had been shown for *Drosophila* germ cell stages. In the mouse, vast differences in mutational response have recently been found between two phases of a single cytologically identified female germ cell stage. The mouse results emphasize the important influence of cellular processes on the mutational outcome and clearly demonstrate that the problem of radiation mutagenesis is far more complex than had been imagined.

Study of the effects of the various biological and radiation factors found to affect mutation frequency in the mouse has led to a number of deductions about the roles of various intracellular and intercellular events in the chain of processes between initial radiation damage and observed mutation. The findings indicate a wide range of mechanisms. At one extreme there is strong evidence, from dose-rate and cell-stage comparisons, for the

* Research sponsored by the U.S. Atomic Energy Commission under contract with the Union Carbide Corporation.

occurrence of a fundamentally important intracellular repair mechanism operating on the initial mutational or premutational damage. At the other extreme, a change in mutational frequency with dose fractionation may be a secondary response to intercellular selection. This paper reviews the more important of these deductions and the experimental findings upon which they are based.

Some of the deductions now appear to be firmly based; others are more speculative and might be regarded as working hypotheses. Although some of the arguments have not been presented before, more of them have. This is a review designed to fit the topic of this symposium by illustrating that mutation as a cellular process can be highly complex in a mammal.

Discussion is limited to induced mutations detected by the specific locus method. The method consists of mating irradiated wild-type animals to animals homozygous for a number of specific visible recessives (seven, in our experiments) and scoring the offspring for mutations at any of these specific, marked loci. The mutations observed in our experiments range from alleles intermediate in effect between wild-type and the marker gene to small deficiencies, too small to be seen cytologically, but detected as deficiencies by genetic tests. The relative frequencies of the different types of mutation occurring under different conditions and in different germ cell stages provide one of the useful tools for analysing the factors affecting the mutation process.

The deductions reached about various links in the chain of processes affecting the observed mutation frequency are listed below as headings for the sections in which the evidence for the conclusion is discussed.

(1) *Repair of mutational or premutational damage can occur under some radiation conditions*

The first evidence that a repair process might be affecting the radiation induction of specific-locus mutations was obtained from dose-rate experiments on the spermatogonial stage in mice. For equal doses, a dose rate of 0·009 R/min gave only approximately one-fourth the mutation frequency obtained with a dose of rate 90 R/min. The possibility existed that this result might not be due to repair, but to a secondary effect of differential killing or other damage in the rapidly dividing and heterogeneous spermatogonial population. However, a similar, but even larger, dose-rate effect was found for irradiated mouse oocytes (Russell, Russell and Kelly, 1958). Since the effect was apparent in the first litter conceived after irradiation, and since these offspring came from non-dividing oocytes of a uniform stage resistant to killing by radiation, the result could not have been due to differential killing. It was concluded that, in oocytes, the dose-rate effect is due to an intracellular repair process. Subsequent evidence supports the view that this conclusion also applies to the dose-rate effect in spermatogonia (Russell, 1961, 1963, 1965*a*).

Further evidence for repair comes from high-dose-rate irradiation given in a small dose, or in a large dose delivered in small fractions. In both cases, the mutation frequencies are significantly less than expected from the frequencies obtained with single large doses at high dose rate (Russell, 1967).

(2) *The dose-rate effect on specific locus mutations is due primarily to repair of "one-hit" rather than of "two-hit" mutational events*

The evidence in support of this conclusion is of two kinds. First, the results of a critical fractionation experiment do not fit the interpretation of reparable two-hit mutations; and second, there is considerable evidence that most of the specific locus mutations, at least in spermatogonia and oocytes, are not two-hit chromosome aberrations.

The fractionation experiment consisted of a comparison of mutation frequencies from high-dose-rate irradiation of oocytes given either as a single 400 R exposure or as a two 200 R exposures spaced 24 hours apart. The observed mutation frequencies were not significantly different. The rationale for the experiment and the conclusion that the results do not fit a two-hit interpretation, regardless of any assumption as to the length of time that breaks remain open, are presented in detail elsewhere (Russell, 1967).

If the reparable specific-locus mutations were predominantly two-hit in type, this would mean that almost all of the mutations produced at high dose and high dose rate would have to be two-hit aberrations. Several pieces of evidence against this have been presented elsewhere (Russell, 1964, 1965*a*). Two additional arguments are given here.

Two of the specific loci used in our experiments, the *d* and *se* loci, are very close together on linkage group 2 (cross-over percentage of only 0·16). In oocytes, about half of the mutations induced at high dose and high dose rate that involve either the *d* or *se* locus also affect the other locus. In other words, half of the mutations at these loci are genetically detected deficiencies. Tests with marker genes close to the *d-se* region show that these detected deficiencies are all small, most of them probably involving less than two cross-over units (L. B. Russell, 1969, personal communication). Since practically all mutations in oocytes are repaired at the lowest dose rate used, the argument that these are predominantly two-hit events would require that most of the mutations at the *d* or *se* loci that do not involve the associated locus should also be two-hit in nature. If these are two-hit deficiencies, they must, on the average, be even smaller than those which affect both loci. It is probably conservative to assume that the total frequency, per average chromosome, of deficiencies as small as those occurring in the *d-se* region is at least 100 times the frequency detected in the *d-se* region. Now, if these small deficiencies are the result of two independent hits occurring close together, the probability of hits occurring farther apart, and causing larger deficiencies, must be much greater. Even if it were only three times greater, the frequency with a dose of 400 R acute X-irradiation to females would, on the above assumptions, amount to more than one large deficiency per genome. Since larger deficiencies are not detected, they must, if they occur at all, be assumed to be lethal either in the germ cells or during development. Only enough oocytes mature in each oestrus to produce the number of eggs ovulated. So an average frequency of at least one lethal deficiency per genome, regardless of whether death occurred in the germ cell or during development, would usually eliminate most of the offspring in the first litter after irradiation. This, however, is not the case: there is only a small reduction in litter size in litters conceived shortly after 400 R irradiation, and most of this may result from single-break aberrations causing dominant lethality.

If the above argument is correct, it would indicate that most of the specific-locus mutations in oocytes are not two-hit deficiencies, and that even the genetically detected deficiencies may be mostly single-hit in origin.

Another new argument against the specific-locus mutations being predominantly two-hit chromosome aberrations comes from current work on chemical mutagenesis in our laboratory. Four different methanesulphonates are being tested both for dominant-lethal and specific-locus mutation induction. All give a dominant-lethal frequency equivalent to that yielded by a large dose of radiation, but only one has given any significant increase over control values for specific-locus mutations, and even there the effect was small. Since there is strong evidence that the dominant lethals are caused by chromosome breakage, the results indicate that chromosome aberrations, including two-hit deficiencies, are probably not the source of most specific-locus mutations.

(3) *The repair process is damaged or saturated at high doses and high dose rates*

There is, so far, no direct biochemical evidence for this conclusion. It is a deduction seemingly necessary in order to account for a dose-rate effect on mutations that appear not to be two-hit events.

(4) *Different germ cell stages, spermatozoa, spermatogonia, and oocytes, have different capacities for repair*

This conclusion is based on the markedly different dose-rate effects on specific-locus mutations obtained in the different germ cell stages. Mutation rate in spermatozoa has, so far, shown no significant dependence on dose rate. The dose-rate effect in spermatogonia is marked over the range from 90 R/min to 0·8 R/min, but further lowering of the dose rate gives no additional reduction in mutation frequency. In contrast to this, oocyte mutation frequency continues to drop to the lowest dose rate tested. At that level, the induced mutation frequency is, so far, not significantly above the control (Russell, 1963, 1964).

These differences provide a striking illustration of the importance of cellular processes on the mutational outcome of exposure to radiation. Lack of repair in spermatozoa might well be expected in view of the extremely low metabolic activity in these atypical cells. Limited repair in spermatogonia may find its explanation in the fact that this cell population is mitotically active and heterogeneous at any one time. There may well be stages in the cycle—for example, just before replication—where repair of damage received at that time is not possible or not efficient. In oocytes, on the other hand, the cell population is in a non-dividing, resting state.

(5) *Different phases within one cytological stage may have markedly different capacities for repair*

One of the most dramatic differences in mutation frequency was found recently to occur between litters conceived shortly after irradiation of the female and those conceived later (Russell, 1965*b*, 1967). Early matings gave appreciable numbers of mutations, while the later matings, in both neutron and X-ray experiments, have given no mutations at all in quite large populations. The offspring from both mating periods come from cells in the same cytological stage, namely, the dictyate oocyte. Of course the early litters come from oocytes that are irradiated in follicle stages that are more mature.

The effect cannot as yet be firmly attributed to differences in capacity for repair. Cell selection or differential sensitivity to initial damage could, theoretically, account for the results. However, a selection mechanism so efficient that every single mutation is eliminated seems unlikely. Some support for the view that differences in repair capacity are involved comes from the work of Oakberg (1967) on [^{3}H]uridine incorporation in mouse oocytes. He showed that the oocytes used in the early post-irradiation conceptions were, at the time of irradiation, in stages corresponding to those in which uridine incorporation has stopped; whereas the stages with zero mutation frequency probably correspond to those which show heavy labelling. Oakberg points out that it is likely that metabolic activity would be closely correlated with capacity for repair.

(6) *The probability of repair appears to be the same at different loci despite large differences in absolute mutation frequency at these loci*

The evidence for this comes from the fact that, in spermatogonia, the relative frequencies of induced mutations at the seven loci remain approximately constant for different dose rates that give a marked change in absolute frequencies (Russell, 1964, 1965*a*). This suggests that all specific-locus mutations induced at high dose and dose rate in spermatogonia are potentially reparable, and that it is primarily a matter of chance which ones are actually repaired at low dose rates.

The fact that virtually all specific-locus mutations, including small deficiencies, seem to be repaired in oocytes exposed to very low dose rates, supports the view that there may be none, or few, that are not potentially reparable in spermatogonia.

(7) *The repair mechanism responsible for the dose-rate effect probably does not involve the repair of a damaged strand by an undamaged one*

The evidence for this comes from the absence of mosaics in the offspring from irradiated spermatozoa. The lack of a dose-rate effect in spermatozoa, and the higher mutation frequency compared with that in spermatogonia, are most plausibly attributed to an absence of repair. If single-strand damage were the kind of damage that failed to repair in spermatozoa, then single-stranded mutations would be induced and would be detectable as mosaics (Russell, 1965*a*).

(8) *There appears to be heterogeneity of mutational response among spermatogonia*

In one of our first radiation experiments with mice we found that the specific-locus mutation frequency in spermatogonia exposed to a dose of 1000 R is less than that from a 600 R exposure (Russell, 1956, 1963). This surprising drop in genetic damage with increasing dose has been attributed to heterogeneity among the spermatogonia for sensitivity to cell killing and, correlated with that, sensitivity to mutation induction. At the higher dose, it is postulated that only those cells survive which are more resistant to killing and which are, at the same time, less susceptible to mutation. The heterogeneity may not involve different types of spermatogonial cells; it could reside simply in the population of one type of cell having, at any one time, a mixture of cells in different stages of the division cycle. This possibility is invoked to explain the results described in the next section.

(9) *There is variation in mutational response during the cell cycle of the spermatogonia*

It was found that division of an acute 100 R X-ray dose into two fractions given 24 hours apart to spermatogonia had a large augmenting effect on mutation frequency, compared with that from a single 1000 R exposure (Russell, 1962). Similar increases in mutation frequency have been found with other fractionated doses spaced at the same 24-hour time interval (Russell, 1965*a*). It is postulated that the cells surviving the first dose were in a radiation-resistant stage of the cell cycle at the time of radiation, and that, 24 hours later, they had progressed to what happens to be a mutationally sensitive stage of the cycle.

(10) *If there is any selective elimination of specific-locus mutations induced in oocytes in mature follicle stages, it cannot be more than a small fraction of the number of mutations observed in the offspring*

The argument is as follows. Let us assume, conservatively, a total of only 5000 loci in the mouse (i.e. an average of only 250 per chromosome) capable of mutating at the average rate observed by us for a 400 R X-ray exposure to oocytes. Alternatively, we could assume 7000 loci mutating at the rate calculated from the combined results of our seven loci and the five different loci studied by Lyon and Morris (1966). In either case, the total mutation rate at this dose would be just over one mutation per germ cell; and, with the above assumptions as to number of loci, this presumably represents a conservative figure. We are supposing that this is the rate for mutations that are not eliminated by selection. Now, if there were just as many more mutations occurring, but not observed because of elimination by selection, then there would be almost no offspring produced. This would be true even if selection occurred among the oocytes, because there are only enough oocytes in mature follicle stage to give one litter. Since there is only about 25 per cent excess death in the first litter after irradiation with 400 R, and since most of this death is attributable to chromosome aberrations (detected by subnuclei), it is clear that there cannot have been extensive elimination of mutations.

(11) *Selective elimination of induced specific-locus mutations cannot be occurring extensively in spermatozoa*

Spermatozoa withstand very high doses without loss of motility or fertilizing capacity, so there is no evidence for any selection occurring before fertilization after a dose of 600 R. Again, assuming only 5000 loci, 600 R would, on the basis of our average per-locus rate for spermatozoa, give more than one mutation per germ cell. The dominant-lethal rate for 600 R is approximately 40 per cent. If all of this were due to specific-locus mutations that were being eliminated, it would represent less than 40 per cent of the mutation rate for mutations that are not eliminated, based on the above conservative assumption as to number of loci. Judging by the frequency of subnuclei in the early cleavage stages, most of the dominant lethals must be due to major chromosome aberrations. So, again, we are able to conclude that relatively few specific-locus mutations can have been eliminated by selection.

(12) *The rarity of deficiencies as detected by the specific-locus method for X- and γ-irradiated spermatogonia is probably not due to selective elimination*

Although deficiencies involving the *d* and *se* loci are rare in X- and γ-irradiated spermatogonia compared with the frequency induced in spermatozoa and oocytes, it would appear that this is not a result of selective elimination. Once induced in spermatozoa, these deficiencies can pass successfully through the spermatogonial stage in the next generation. Another convincing piece of evidence is that *d-se* deficiencies are recovered at an appreciable rate from spermatogonia irradiated with neutrons. When they *are* induced they can get through to the offspring.

SUMMARY

Analysis of the effects of the various biological and radiation factors found to affect mutation frequency in the mouse has led to a number of deductions about the roles of various intracellular and intercellular events in the chain of processes between initial radiation damage and observed mutations scored by the specific-locus method.

Repair of mutational or premutational damage can occur at low radiation dose rates and at low doses. This is due primarily to repair of "one-hit" rather than of "two-hit" mutational events, but probably does not involve repair of a damaged strand by an undamaged one. Spermatozoa, spermatogonia and oocytes have different capacities for repair. Even different phases within one oocyte stage may have markedly different repair capacities. Most mutations occurring in oocytes and spermatogonia appear to be potentially reparable. There is heterogeneity of mutational response during the cell cycle of the spermatogonia. In all germ cell stages for which there is any evidence, there appears to be little selective elimination of induced specific-locus mutations.

REFERENCES

LYON, M., and MORRIS, T. (1966). *Genet. Res.*, **7**, 12.
OAKBERG, E. F. (1967). *Archs Anat. microsc.*, **56**, 171.
RUSSELL, W. L. (1956). *Genetics, Princeton*, **41**, 658.
RUSSELL, W. L. (1961). *J. cell. comp. Physiol.*, **58**, Suppl. 1, 183.
RUSSELL, W. L. (1962). *Proc. natn. Acad. Sci. U.S.A.*, **48**, 1724.
RUSSELL, W. L. (1963). In *Repair from Genetic Radiation Damage*, pp. 205–217; 231–235, ed. Sobels, F. Oxford: Pergamon Press.
RUSSELL, W. L. (1964). In *Genetics Today*, pp. 257–264, ed. Geerts, S. J. Oxford: Pergamon Press.
RUSSELL, W. L. (1965*a*). *Jap. J. Genet.*, **40**, Suppl., 119–127.
RUSSELL, W. L. (1965*b*). *Proc. natn. Acad. Sci. U.S.A.*, **54**, 1552–1557.
RUSSELL, W. L. (1967). *Brookhaven Symp. Biol.*, **20**, 179–189.
RUSSELL, W. L., RUSSELL, L. B., and KELLY, E. M. (1958). *Science*, **128**, 1546–1550.
United Nations Scientific Committee on the Effects of Atomic Radiation (1958). Report to the General Assembly. Official Records: Thirteenth Session, Supplement No. 17 (A/3838). New York: United Nations.

DISCUSSION

Magni: Is the phenomenon you observed really representative of what could happen to any type of mutation that can theoretically occur in those genes, Dr. Russell? Your system for detecting the mutations doesn't select a certain type of mutation, does it? It appears to me that you can detect only mutations which are not complementing in the diploid stage with the original mutant. The new complementing mutation will be completely lost. As complementing mutants are base substitutions, in general you might be selecting for more drastic alterations in the chromosome and completely losing base substitutions in the genes. In this case it could be that repair phenomena could hold true for small losses within the gene but not for minor changes like base substitutions.

Russell: We have to limit our conclusions, of course, to the types of mutation that we are seeing. We could be failing to detect certain types of mutation. We can say, however, that we are not missing many lethal mutations. As I pointed out in my paper, the total mutation frequency of events that are lethal cannot be much higher than the frequency of the mutations that we detect.

Auerbach: Did you mean your suggestion could explain the fractionation effect, or dose-rate effect, as a two-hit phenomenon, Professor Magni?

Magni: I was just wondering whether a repair phenomenon holds true for any kind of genetic lesion which can be induced in a gene, or only for a fraction of them, let's say deletions of short sequences.

Kimball: I have to challenge the idea that base substitution mutations at all loci are necessarily going to lead to complementing situations. For example, in the adenine-3A loci in *Neurospora* there is no intra-allelic complementation yet they are perfectly good base substitution mutants. Intra-allelic complementation is only seen in certain special situations. Nothing of this sort happens in adenine-3A although it happens in adenine-3B.

Magni: My question should have been preceded by another: are any of these loci complex loci?

Apirion: I don't think that the problem of intragenic complementation is serious, since even in complementing loci often the number of mis-sense mutations which do complement is not too high.

Magni: I must challenge this statement. In *Saccharomyces cerevisiae* the great majority of mis-sense base substitutions in the arginine-4 locus are complementing each other in a complete fashion.

Russell: We are fully aware that the type of repair process we are exploring is being measured by mutations detected by the specific-locus method. We have recently started experiments with an entirely different method, namely XO induction; that is, X-chromosome loss presumably resulting from chromosome breakage. This also appears to show a dose-rate effect. So there is evidence for repair of this type of damage too. However, so far, there has been no evidence for repair of XO induction at low doses. More work is needed to explore this complication.

Clarke: Is there any evidence that highly mutable or mutation-refractory strains of mice exist? In other words, do you get genetic background effects on the ionizing mutability?

Russell: We have not done extensive tests with other strains. Most of our work has been with a hybrid between the 101 and C3H inbred strains. We have done some mutational work on the C3H strain, and so far this behaves the same way as the hybrid. However, in recent work we have seen marked strain differences in response to chemical mutagens. This could be due simply to differences in physiological response to the mutagen, for example, whether the chemical gets through to the germ cells or not. We are now testing this.

Sobels: I have some difficulty in reconciling your interpretation in terms of repair of premutational damage with the findings after exposure to fast neutrons. With neutron irradiation, high relative biological effectiveness (RBE) values of 5 or 6 are observed for specific-locus mutations in both spermatogonia and oocytes. A similar RBE value is found for dominant lethals induced in mouse spermatozoa. Since we may suppose that the dominant lethals reflect chromosome breakage phenomena, and since in other organisms, as *Drosophila*, the RBE for point mutations is considerably smaller, the neutron data would suggest a sizeable two-hit component among the total yield of specific-locus mutations.

Russell: The neutron data do show a much higher frequency of detectable chromosomal damage. If we simply lump together all specific-locus mutations, the neutron data include a larger component of detected chromosomal aberrations. For example, we get a higher frequency of *d-se* deficiencies induced in spermatogonia with neutrons than with X- or γ-rays.

Sobels: Do you mean that the pattern of mutations induced by neutrons is different from that recovered after X-irradiation?

Russell: The evidence for repair is also quite different with neutrons. We have some evidence for a dose-rate effect of neutrons in females, but so far none at all for neutron irradiation of spermatogonia. On the other hand, the effect on mutation frequency of the interval between irradiation and conception in females treated with neutrons is just as great as in females exposed to X-rays. In both cases, the mutation frequency in later litters drops to very low levels. This result also suggests, among other possibilities, that repair may occur with neutrons. This is perhaps not as unusual as many people believe. Dr. Kimball has strong evidence that repair is possible with densely ionizing particles.

Kimball: With α-particle irradiation and recessive lethal mutations in *Paramecium* one can (Kimball, Gaither and Wilson, 1959), by post-treatment with streptomycin, which has been one of our standard procedures, very appreciably reduce the yield of mutation. In fact the amount of the reduction was not significantly different from that found with X-rays. One certainly can, by a post-treatment condition which does not involve cell selection or anything of this sort, reduce the yield of mutation produced by high linear energy transfer radiation.

Sobels: Is it not possible that this affects specifically the two-hit component of the total mutational damage?

Kimball: The two-hit component with X-rays is certainly very small. I can't speak about the two-hit component with α particles, except that I wouldn't have expected one from anything known in other organisms.

Auerbach: I am always sceptical about dominant lethals as indicators of chromosome breakage. What about translocations?

Russell: Translocations have been found after treatment with ethyl methanesulphonate (Cattanach, Pollard and Isaacson, 1968).

Auerbach: Although X-ray-induced dominant lethals probably are due to chromosome breakage one cannot say the opposite, because all kinds of effects on the sperm may prevent it from initiating development.

Russell: These dominant lethals are detected by embryo loss in both preimplantation and post-implantation stages. They have been checked in early cleavage stages. Subnuclei are occurring in the early cleavage cells, which I think is strong evidence of chromosome breakage.

Loveless: Which four methanesulphonates produced dominant lethals?

Russell: Methyl, ethyl, propyl and isopropyl.

Sobels: Repair in oocytes is measured by two different criteria. In the late oocyte you postulate that repair of premutational damage occurs on the basis of the effects of dose rate, low doses *per se* and dose fractionation. In the early oocytes, the absence of mutations from cells sampled later than seven weeks after radiation exposure is taken as an indication for repair. Is it not possible that this interval effect, which is also observed after neutron irradiation, does not reflect repair, but selective killing of cells with mutations?

Russell: I have always entertained this possibility, but, as I have explained elsewhere, repair seems to be a more plausible hypothesis (Russell, 1968).

Sobels: How much does the dose-rate effect with neutrons reduce the mutation frequency in the late oocyte stages?

Russell: We get approximately a halving of the mutation rate, which is statistically significant at about the 1 per cent level.

Kimball: With α particles in *Paramecium* in the G2 period it is difficult to be sure that there is any mutation induction at all (Kimball, 1961). This is essentially in line with your observation, Dr. Russell, that in early oocytes it is difficult to produce any mutation.

Sobels: The low dose of 50 R produces a significantly lower mutation frequency than you would expect by extrapolation from the results with 400 R. Do you get a better fit to a linear than to a dose-squared relationship? Assuming for a moment that there is a two-hit component among your total mutational damage, the concave dose-square curve would accommodate the lower mutagenic effectiveness at low dose levels.

Russell: The fact that the curve is concave upwards in the range 0–200 R will fit both a one-hit and a two-hit interpretation of the mutational damage. Those data are not conclusive. The most critical single piece of evidence against the damage being two-hit is the fractionation experiment. Two fractions of 200 R spaced 24 hours apart give no reduction in mutation frequency compared with that from a single 400 R exposure (Russell, 1968).

Sobels: What is the interval between the fractions when a reduction of the mutation frequency is observed after a dose of 400 R is split into eight fractions?

Russell: Fractions were given $1\frac{1}{4}$ hours apart. Repair can apparently occur when fractions of 50 R are spread out over this length of time.

REFERENCES

CATTANACH, B. M., POLLARD, C. E., and ISAACSON, J. H. (1968). *Mutation Res.*, **6**, 297–307.
KIMBALL, R. F. (1961). *J. cell. comp. Physiol.*, **58**, suppl. 1, 163–170.
KIMBALL, R. F., GAITHER, N., and WILSON, S. M. (1959). *Radiat. Res.*, **10**, 490–497.
RUSSELL, W. L. (1968). *Brookhaven Symp. Biol.*, **20**, 179–189.

FINAL DISCUSSION

Auerbach: Discussions of mosaics become confused because we all talk about different things. First we have mosaics in higher organisms as found cytologically. Here the distinction between chromosome and chromatid breaks plays a great role in the discussion of chemical mutagens versus X-rays. Chemical mutagens almost always produce chromatid breaks, even when treatment has been given in G1. Possibly the mosaics that arise in this way have a completely different origin from the mosaics in micro-organisms.

Dr. Slizynska (1969) in our Institute has worked on *Drosophila* chromosomes, which are just as respectable as plant chromosomes and probably respond to treatment in the same way. She analysed cytologically the F1 larvae from chemically treated males by looking at the salivary gland chromosomes. As with plant chromosomes, she found a preponderance of mosaic rearrangements. Her conclusion about their origin is derived from an experiment in which she treated with one of those mutagens with which storage effects occur. After treatment of spermatozoa with triethylene melamine, chromosome breaks started accumulating during the storage of treated chromosomes in the seminal receptacles of untreated females. Chromosomes treated with triethylene melamine had many more mosaics than those treated with X-rays when she examined them immediately after treatment. However, after storage for about a week or nine days the spectrum of rearrangements had become very similar to that induced by X-rays. From this, and from the special types of rearrangement that she got, she derived the following explanation. In the still undivided two-strand structure, the chemical produces potential breaks which occur at the same sites in both strands. Mosaics may arise when these break open after the chromosome has replicated, because then it is possible for some breaks to rejoin in one way and for others to rejoin in another way or not to rejoin at all. With two breaks, for example, a piece from one chromatid may become inserted into one of the breaks of the other chromatid. This results in a tandem or reverse duplication in one chromatid, and a deletion in the other. After X-ray treatment, the breaks probably open almost at once, so one doesn't get this situation. However, with triethylene melamine, and probably with most chemicals, there is a delay in the opening of breaks and it is this that leads to mosaicism. So these mosaics start really as potentially complete changes and end up as mosaics because the opening of the break has been delayed until there are two structures which can handle the breaks differentially. Whether the same happens in plant chromosomes I don't know; it is difficult to test, because I don't know of any system where one could do what Slizynska did, that is, store treated spermatozoa without replication.

Sobels: The replication has nothing to do with the storage, has it?

Auerbach: No.

Secondly we have mosaics for gene mutations. Professor Sobels probably wants me to say something about mosaics in *Drosophila*, which have been studied quite extensively, but the difficulty there is that one does not really know what types of changes occur. One gets a mixture of chromosome rearrangements and gene mutations. I want to discuss what are probably real gene mutations. These are mutations from red to white in fission yeast. Here, mosaics can easily be seen as colonies that are part-red, part-white. In the interpretation confusion arises because no distinction is made between two completely different causes of mosaicism. The first is damage to only one strand of DNA. This will lead to half-and-half mosaics. Dr. Nasim in our unit (Nasim and Clarke, 1965) found that the mode of the distribution of sector sizes among nitrous acid-induced mosaics was about 50 per cent mutated cells. But there is also a completely different source of mosaicism: instabilities *which replicate as such*. We don't know their nature. They may affect both strands or one strand. If one just scores mosaics and does not go on to further generations one cannot distinguish between the two types of mosaic.

Various suggestions have been made as to how one might account for "completes" since, after all, DNA is double-stranded and most mutagens probably act only on single strands. I don't really see any necessity for invoking the master-strand hypothesis because, if one does use it, one has to find an additional hypothesis to explain the mosaics. Vielmetter (Schuster and Vielmetter, 1960) put forward the so-called lethal-hit hypothesis: if there is a mutation in one strand and if the same mutagen produces a lethal hit on the other strand, then the second strand will be lost, and this will give a completely mutant clone. Vielmetter's data were not really in accordance with this hypothesis. He showed that nitrous acid produced mottled plaques in a double-stranded phage, and only complete plaques on phage φX 174; he pointed this out himself. If the lethal-hit theory is right, then as one increases the dose of the mutagen, so should the number of lethal hits increase and the frequency of mosaics decrease. However, in his experiments the frequency of mosaics was more or less dose-independent. In our system, several—although not all—mutagens yielded fewer mosaics with increasing dose and this may well have been due to an increased frequency of lethal hits. The question we asked ourselves was whether lethal hits can explain *all* "completes"; for this, one has to look at the data quantitatively. We made the—to me—plausible assumption that a dose that produces twice as many "mutational hits" will also produce twice as many "lethal hits". This allowed us to predict how the frequency of "completes" should change with dose, if lethal hits were the only source of "completes". For none of the mutagens did the data agree quantitatively (Nasim and Auerbach, 1967). In summary, there is no doubt that lethal hits, where they do occur, will contribute to the "completes". But there *is* doubt, from our experiments, whether this is the only source of "completes". If it is not the only source, as we think, then we have to look for some other explanation.

There are two theories, either of which could account quantitatively for our results. One is that the "completes" are due to some mechanism different from that causing mutation, for instance to cross-linkage. The other is that "completes" arise from mosaics by a repair mechanism which monitors the fit of the two strands. For instance, if there has been a change from A to G then the repair enzyme will either reverse it, in which case the mutation is lost, or it will change the T on the opposite strand into C, in which case the mutation becomes "complete".

This is the repair hypothesis, for which at the moment there is no proof, but I think there is also no proof against it. People have looked for mosaics in repairless systems and if they don't get a difference, they say this is against the repair hypothesis. I think it is neither for nor against, because there are many repair systems and one doesn't know which may be involved.

We have started to test the repair hypothesis in a completely different way, by using a combination of mutagens that produce different proportions of mosaics and completes. The argument runs like this. Hydroxylamine produces very many mosaics; u.v. produces far fewer. If this is so because the inhibition of DNA synthesis by u.v. allows more time for repair, then one might expect that a following dose of hydroxylamine would produce fewer mosaics than it would do otherwise. Curiously enough, exactly this experiment has been done by Freese (Bautz Freese and Freese, 1966) with the result that I predicted. He interpreted it in a different way, which is legitimate for phage but which would hardly be legitimate for yeast. If I should get the same result as Freese, I would take it in favour of the repair hypothesis.

Now let me come to the instabilities. Usually when I say something about the origin of replicating instabilities, someone mentions the experiments by Krieg (Green and Krieg, 1961). He treated phage with ethyl methylsulphonate and found that mutations continued to occur at the same rate at every replication up to the time of lysis. He assumed that this was due to the persistence of the alkylated or depurinated template, which at every replication had the same probability of causing an error leading to mutation. This is probably the correct interpretation for Krieg's case, but it cannot account for cases in which an instability replicates *as instability*, because neither an alkylated guanine nor an apurinic gap would be expected to replicate themselves. Yet this is just what happens when *one* mosaic, presumably carrying a newly induced instability, produces *several* mosaics for the same mutant gene. I found this many years ago after chemical treatment of *Drosophila*, and now we and others have found it again in *Drosophila* as well as fission yeast.

Evans: Are you convinced that the mosaics that result from a replicating instability are not produced as a result of a deficiency?

Auerbach: I used to think they might be duplications that give rise to deficiencies. A duplication, which is easily produced by chemicals, might be a source of a replicating instability, because at replication it might fold back on itself and thus might be missed out. This is how Demerec (Itikawa and Demerec, 1967) explained what I would call a replicating instability in bacteria. So far, our attempts to correlate replicating instabilities in *Drosophila* with duplications have not been successful, but we have only looked at very few.

One system where the source of a replicating instability is a chromosome rearrangement is being studied by Nga and Roper (1966). They feed translocations into a strain of *Aspergillus* and then let them segregate out. This results in duplications, which are sources of replicating instabilities. But I don't think that this can apply to our instabilities in yeast. Dr. Nasim (1967) found that out of five loci that could be involved in yielding the white sector, only one had become unstable in any particular line, but different ones in different lines. Workers in Pisa (Abbondandolo *et al.*, 1967; Loprieno *et al.*, 1968) have looked at a different system of mutations, also in fission yeast. By means of intragenic recombination experiments, they found that not only the same locus but the same site was involved in any particular line, and not the same in all of them. This makes it very unlikely that these instabilities are due to chromosomal rearrangements. Finally, there are the instabilities which Professor Dawson and Dr. Smith-Keary (1963) discovered in *Salmonella* and which might be due to episomes.

Kimball: There is trouble with the repair hypothesis. It necessitates that the mutation be fixed at the time of repair, in some sense, because the whole hypothesis depends on putting in the wrong bases at the time of repair. It seems to me there is fairly good evidence from Witkin's work (1967) on the Rec⁻ mutations, with and without excision repair operating, that considerable amounts of repair can go on with no mutation production whatsoever. Also some of our data with *Paramecium* (Kimball, 1966) suggest rather strongly that the major source of mutation is not fixation at the time of repair but fixation at or near the time of replication. Yet mutations in this organism are primarily complete and not partial.

Auerbach: We don't know whether dark repair occurs in *Schizosaccharomyces.*

Clarke: We do know that dark repair occurs in *Schizosaccharomyces pombe* u.v. mutants.

Kimball: I think we have to accept in any repair hypothesis that mutation is fixed at the time of repair. It seems to me that this is essential to the whole thing. If there is little or no fixation of mutation at the time of repair then I have great problems with the hypothesis.

Auerbach: What do you mean by fixation? Do you mean insertion of a mutated base?

Kimball: In the repair hypothesis one puts in the wrong base. Once you have got that far the mutation is there in the replicating form.

Maaløe: In later studies Vielmetter, Messer and Schütte (1968) observed systematic changes in the ratio of solid versus solid+sectored colonies as a function of the map position of the character they looked at. This does not fit the lethal-hit notion, in the sense that if one considers a locus close to the origin of replication most of the cells in the population will carry two copies of this locus and *vice versa* for positions close to the end of the replicating unit. From this one can work out an expectancy for sectored colonies among all colonies at any time subsequent to the mutagenic event.

Kubitschek: It seems to me that one can explain mosaicism on a master strand model by assuming that repair occurs in some subsequent generation. We know that repair can occur after the first division from the work of Bridges and Munson (1968) and from Sauerbier's work (Sauerbier and Hirsch-Kauffmann, 1968) with phage, for example, where thymine dimers are carried through to the progeny.

Auerbach: If I have to make assumptions I prefer to make one rather than two. I have only to assume repair and you have to assume both master strand and repair.

Sobels: Would you predict that the lethal hits have a higher dose exponent than the mutational damage?

Auerbach: I think that dose-effect curves for mutations in cellular organisms are composite curves, and that the dose exponents must be interpreted with great caution.

Sobels: In *Drosophila* one finds more completes at higher doses (Inagaki and Nakao, 1966).

Auerbach: This may well be due to lethal hits, which certainly contribute to the completes. The only question is whether they can *quantitatively* account for them and for our data they can't. Apparently they can for Vielmetter's data.

Sobels: What seems worth doing is to test whether a change occurs in the ratio of completes to mosaics in a system where straightforward repair can be measured.

Auerbach: The trouble is that there are so many repair systems. Nasim tested two strains of *Saccharomyces* which are more u.v. sensitive than the wild type. They did produce more mosaics but the effect is not very striking.

Sobels: How does the phenomenon of lethal sectoring which has been studied so extensively in yeast fit in with your ideas?

Auerbach: That is the converse. A lethal sector is presumed to represent a lethal hit and therefore should lead to a complete mutation. Haefner (1967) did a lot of this work, but we agreed that lethal sectors cannot explain all "completes" in fission yeast, although there is no doubt that they contribute to them.

Dawson: In only two of the five systems of instability that we have looked at in *Salmonella* is there evidence which points in the direction of what we call controlling episomes. In the other three we have no evidence that the instability has a mechanism of that type. I am very conscious, in much of the discussion about replicating instabilities, of the inadequacy of the evidence for distinguishing between possible mechanisms. For example is it known, for the instabilities that are induced in *Drosophila* by chemical mutagens, whether the instability is even inherited at the gene whose instability you are scoring? May it not be, as in maize (McClintock, 1965) and delphinium (Dawson, 1964), that the instability is a two-element system and you have induced something at the regulator rather than at the gene whose phenotype you are scoring? Also one very seldom seems to know whether the instability is referrable to any particular site in the gene, or whether it is an instability over considerable regions of the genome. In much of the work that one hears discussed—mosaic colonies and so on—

even the evidence as to whether the instability is in both directions or not is often missing.

Auerbach: I realize this. That is why Nasim is now making a great effort to obtain such evidence. The difficulty with these mosaic yeast colonies is to be absolutely sure that even a small sector of one type may not contain one or a few cells of the other. Dr. Nasim told me that he has done the testing very carefully and over several generations, and that there are some cases where both the red and the white sectors appear to be unstable. I think, however, that in yeast the duplication idea can be excluded, at least for those cases where the instability is site-specific.

Sobels: Fox was able to show that acridines can effectively remove the postulated "exosome". Fox (1968) treats with DNA carrying specific marker genes. He then observes replicated instabilities for the markers concerned. These are transmitted and segregate in a purely Mendelian fashion, and the instability can persist for as many as 12 generations. Fox explains his observations on the basis of an episome model. The episome is replicated in close association with the chromosome, but never becomes incorporated, and is therefore called an "exosome". There is thus a variation of the expression, but not of the transmissibility.

Auerbach: I can't give an answer to this. All I wanted to do was to clear up the confusion about mosaics. One is dealing with very different phenomena and these should be kept apart. An explanation that fits one situation does not necessarily fit the other.

Magni: One must be sure that sectoring colonies are really segregating the two phenotypes and do not derive from two stable cells which give rise to different proportions of cells in the colony.

Auerbach: Nasim (1967) did a very simple test which ruled out this source of error. He used three unstable lines for which he knew which one of the five possible loci had become unstable. He then mixed the unstable red cells of each line with stable white cells representing mutation at a different locus. He plated the mixtures, looked for mosaics and tested them. In every case, the mosaics carried mutations at the unstable locus characteristic for the line and no mosaics had been formed with the admixed cells.

Kaplan: I studied inheritable instabilities many years ago in *Serratia* where colour mutations appear at very high frequencies after u.v. irradiation (Kaplan, 1959). Most of the mutant colonies are totally changed—they are white or pink in different degrees—but some are sectored. When post-cultures of these sectored colonies are made one usually gets different colour types from one colony—in addition to sectored colonies; again there were also whites, and pinks of several types, and often the wild type.

Winkler (1960) could distinguish four groups of whites by cross-feeding, as if they represented different cistrons for pigmentation. When whites from different sectored colonies were tested there was always only one type in a colony. That means that only one particular colour gene was mutated in the mother cell of the colony. Our interpretation is that the mutation to a very unstable allele of a colour gene is induced by u.v. in the cell growing to a sector colony—unstable in the sense that it mutates spontaneously and frequently to other alleles during the growth of the colony, thus giving the sectors.

Auerbach: The question is why is it unstable?

Kaplan: I think because the allele is highly mutable spontaneously. When one induces auxotrophic mutations by a mutagen one gets also a broad scale of mutants of very different spontaneous back mutabilities, e.g. from 10^{-3} to 10^{-9}. I think those unstable alleles of pigment genes are nothing but the most extreme part of such a range of mutabilities; at the other end of the scale are the more or less stable colour mutants which are also found.

Auerbach: In fission yeast the frequency with which instabilities are produced depends on the mutagen. One gets few instabilities with hydroxylamine and nitrous acid, although it produces many mosaics by one-strand hits; one gets many instabilities with ethyl methanesulphonate.

Kaplan: But isn't it possible that different alleles are preferentially induced by these different mutagens?

Clarke: The evidence that replicating instabilities occur in *S. pombe* seems to me unsatisfactory, in the absence of single cell pedigree analyses by micromanipulation techniques.

Auerbach: I know that when you looked at mosaics, Dr. Clarke, you were rightly worried about the possibility that spurious mosaics may arise through clumping, as Professor Magni outlined. However, this source of error has now been ruled out for our system by Nasim's experiments (1967) which showed locus specificity of the instability, and for the system used in Pisa by experiments (Abbondandolo *et al.*, 1967; Loprieno *et al.*, 1968) that showed site specificity. Simultaneously, these findings dispose of the assumption that instabilities are due to the induction of mutator genes by the treatments used.

Dawson: In *Salmonella* Dr. Ryce (see Riyasaty and Dawson, 1967) has used a particular allele of a tryptophan A gene to look for further mutations to auxotrophy. She finds a disproportionally large number of tryptophan A auxotrophs. The first thought was that this was instability at the tryptophan A gene, somehow induced by this mutant site at tryptophan A. But when one isolates those genotypes which don't have the original mutant site but have the additional mutations that have occurred in tryptophan A, they are not auxotrophic. In other words one has simply changed a point in the gene such that many other mutations are now expressed phenotypically as mutants, whereas if one didn't have that initial change in the gene one would never have scored them. That is at least a possible interpretation and is consistent with the data. In other words, notwithstanding the observation of the burst, if you like, of the large number of tryptophan A auxotrophs, it does not necessarily have anything to do with high rates of mutation.

Apirion: The hypothesis may perhaps be correct for replication after mutagenesis but not for normal replication. I don't like this kind of hypothesis, but an alternative is that there may be excision systems in the cell, not just for thymine dimers but for many other altered bases. If such a system exists then of course only one of the strands would be copied.

Auerbach: This is the repair hypothesis.

REFERENCES

ABBONDANDOLO, A., BONATTI, S., GUGLIELMINETTI, R., and LOPRIENO, N. (1967). *Microb. Genet. Bull.*, **26**, 1.

BAUTZ FREESE, E., and FREESE, E. (1966). *Genetics, Princeton*, **54**, 1055.

BRIDGES, B. A., and MUNSON, R. J. (1968). *Biophys. biochem. Res. Commun.*, **30**, 620.

DAWSON, G. W. P. (1964). *Genet. Res.*, **5**, 423–431.

DAWSON, G. W. P., and SMITH-KEARY, P. F. (1963). *Heredity, Lond.*, **18**, 1.

FOX, A. (1968). *XII Int. Congr. Genet.*, Tokyo.

GREEN, D. M., and KRIEG, D. R. (1961). *Proc. natn. Acad. Sci. U.S.A.*, **47**, 64–72.

HAEFNER, K. (1967). *Genetics, Princeton*, **57**, 169.

INAGAKI, E., and NAKAO, Y. (1966). *Mutation Res.*, **3**, 268–272.

ITIKAWA, H., and DEMEREC, M. (1967). *Genetics, Princeton*, **55**, 63.

KAPLAN, R. W. (1959). *Arch. Mikrobiol.*, **32**, 138–160.

KIMBALL, R. F. (1966). *Adv. Radiat. Biol.*, **2**, 135.

LOPRIENO, N., ABBONDANDOLO, A., BONATTI, S., and GUGLIELMINETTI, R. (1968). *Genet. Res.*, **12**, 15.

MCCLINTOCK, B. (1965). *Brookhaven Symp. Biol.*, **18**, 162–182.

NASIM, A. (1967). *Mutation Res.*, **4**, 753.

NASIM, A., and AUERBACH, C. (1967). *Mutation Res.*, **4**, 1.

NASIM, A., and CLARKE, C. H. (1965). *Mutation Res.*, **2**, 395.

NGA, B. H. G., and ROPER, I. A. (1966). *Heredity, Lond.*, **21**, 530.

RIYASATY, S., and DAWSON, G. W. P. (1967). *Genet. Res.*, **10**, 127–134.

SAUERBIER, W., and HIRSCH-KAUFFMANN, M. (1968). *Biophys. biochem. Res. Commun.*, **33**, 32.

SCHUSTER, H., and VIELMETTER, W. (1960). *J. Chim. Phys.*, **58**, 1005.

SLIZYNSKA, H. (1969). *Mutation Res.*, **8**, 165–175.

VIELMETTER, W., MESSER, W., and SCHÜTTE, A. (1968). *Cold Spring Harb. Symp. quant. Biol.*, **33**, 585–598.

WINKLER, U. (1960). *Z. allg. Mikrobiol.*, **1**, 83–85.

WITKIN, E. M. (1967). *Brookhaven Symp. Biol.*, **20**, 17–55.

INDEX OF AUTHORS*

Numbers in bold type indicate a contribution in the form of a paper; numbers in plain type refer to contributions to the discussions.

*Author and Subject Indexes compiled by Mr. William Hill.

INDEX OF SUBJECTS

Printed by Spottiswoode, Ballantyne & Co. Ltd., London and Colchester